Hong Kong Landscapes

Hong Kong Landscapes
Shaping the Barren Rock

Bernie Owen and Raynor Shaw

Hong Kong University Press
14/F Hing Wai Centre
7 Tin Wan Praya Road
Aberdeen
Hong Kong

© Hong Kong University Press 2007

ISBN 978-962-209-847-3

All rights reserved. No part of this publication may be reproduced or transmitted, in any means, electronic or mechanical, including photocopy, recording, or any information storage or retrieval system, without prior permission in writing from the publisher.

British Library Cataloguing-in-Publication Data
A catalogue record for this book is available from the British Library.

Secure On-line Ordering
http://www.hkupress.org

Cover photograph: Easterly view of Lantau Peak from Por Kai Shan, Lantau Island
Title page photograph: Ngong Ping Plateau, Ma On Shan Country Park

Printed and bound by Colorprint Production Co. Ltd. in Hong Kong, China

Lion Rock from the MacLehose Trail

Table of Contents

List of Information Notes... viii
Using this Book... ix

Part One: Introduction
Modern and Ancient Environments... 2
   Preamble 2
   The Lay of the Land 2
   Reconstructing the Past 5

An Environmental History 7

Part Two: Landscape Types and Origins
Rugged Mountain Landscapes: A Story of Ancient Volcanic Eruptions 13

Rounded Hilly Landscapes: The Roots of Volcanoes 21

Ridges and Colourful Landscapes: Ancient Seas, Rivers, Lakes, and Deserts 25

Lowlands and Valleys: Fractured Rocks and Rivers 33

Coastal Landscapes: Cliffs, Beaches, and Mud Flats 41
   The Western Region 41
   The Central Region 43
   The Eastern Region 45

Part Three: Hong Kong Regions
The Northwestern New Territories... 48
   What Lies Beneath 50
   Rivers and Floodplains 54
   Human Impacts 60

The Northeastern New Territories 67
   Landscape Foundations 68
   Streams and Waterfalls 71
   Marine Inlets and Islands 74
   Human Impacts 77

The Western New Territories 84
   Magma Chambers and Eruptions 86
   High Ground, Low Ground 88
   Gullies and Badlands 90
   Mountain Streams 94
   Human Impacts 96

The Central New Territories 101
    An Era of Violent Eruptions   104
    Human Impacts   109

The Southeastern New Territories 119
    Geological Background   121
    Sea Level Change and Islands   125
    Human Impacts   127

The Eastern New Territories 131
    Rocks and Scenery   133
    Coastal Environments   137
    Human Impacts   145

Lantau Island 149
    Ancient Sediments, a Caldera, and Multiple Intrusions   152
    Mass Movements and Landslides   158
    Coastal Landscapes   160
    Human Impacts   164
    Expanding Infrastructure   169

Kowloon and the Lion Rock Ridge 173
    Geology and Weathering   175
    Human Impacts   180

Hong Kong Island and Lamma 189
    Landscape Foundations   192
    Human Impacts   195

Seas and Islands 207
    Marine Environments   208
    Human Impacts   214
    The Islands of Hong Kong   215

Epilogue: Landscapes, Past, Present, and Future 230
    Global Warming   232
    Urbanisation and Landscape   233

Information Sources and Further Reading 235
    Part I: Publications Referred to Extensively   235
    Part II: Publications Referred to Less Extensively   235
    Part III: Further Reading   239

Index 242

List of Information Notes

01	Geological Time and Dating	6
02	Extrusive Igneous Rocks	15
03	Escarpments and Cuestas	18
04	Intrusive Igneous Rocks	22
05	Beds, Bedding Planes, and Time	26
06	Sedimentary Formations	29
07	Differential Erosion	31
08	Types of Fault	35
09	Joints in Rock	37
10	Boulder Fields and Joints	39
11	Metamorphic Rocks	51
12	Folds	52
13	Mai Po and the Wetlands	55
14	Floodplain Sediments	57
15	Feng Shui and Landscape	83
16	Plate Tectonics	88
17	Colluvium	91
18	Granite Weathering	92
19	Base Level and Streams	95
20	Plantation Trees	98
21	Hong Kong Minerals	100
22	The Ng Tung Chai Waterfalls	102
23	Migrating Volcanoes	105
24	Earthquake Risks in Hong Kong	108
25	Tea Terraces	109
26	Animal Biodiversity	110
27	Reservoirs	111
28	Country Parks	112
29	Mining and Landscape Scars	114
30	The New Town Programme	115
31	Tai Po Kau	116
32	Scheduled Areas	118
33	The Ma On Shan Iron Mine	124
34	Valley Deltas and Progradation	126
35	Chinese Graves	128
36	Landfills	129
37	Butterflies	130
38	Hong Kong Weather	132
39	Insects	134
40	Calderas and Columnar Joints	136
41	Beach Terminology	139
42	Beach Processes	141
43	Delta Sediments	143
44	Hong Kong Corals	145
45	The Lantau Dyke Swarm	155
46	Mass Movements	159
47	Tidal Flats	162
48	Current Ripples	163
49	Pirates and Forts	165
50	Dolphin Habitat Problems	172
51	Weathered Rock Layers	177, 178
52	Urban Geological Mapping	181
53	Urban Geology	183
54	World War II	187
55	Amphibians and Reptiles	206
56	Tides	210
57	Seismic Profiles	213
58	W. Brother Isl. Graphite Mine	216
59	Vietnamese Refugees	217

Using This Book

This book is divided into three parts. Part One (Introduction) outlines the main objectives of the book, and describes the range of modern environments that have developed in Hong Kong. This section also provides a brief environmental history of Hong Kong over a geological time scale that covers the last 400 million years.

Part Two (Landscape Types and Origins) identifies five major landscape types: rugged mountains; rounded hilly landscapes; ridges and colourful landscapes; lowlands and valleys; and coasts. Each type of landscape is discussed in a separate chapter, which focuses on the major geological factors and surface processes that have influenced the formation of these contrasting settings. In these chapters basic concepts are explained that are applied in later sections of the book.

In Part Three (Hong Kong Regions), Hong Kong is subdivided into ten regions. Each region is given a separate chapter in which the processes that shaped that region's landscapes are explained. Each chapter contains an introduction, and a description of the geology and its influence on landscape. This is followed by a discussion of surface processes and the human factors that contribute to landscape development.

The book presents basic concepts first. These are later built on as the story of landscape origins becomes more complex. Each chapter in Part Three emphasises a different aspect of landscape. Detailed background information is provided in numbered Information Notes in blue boxes. These are referred to in the text, for example, as: IN15, together with a page number if the note is on a different page.

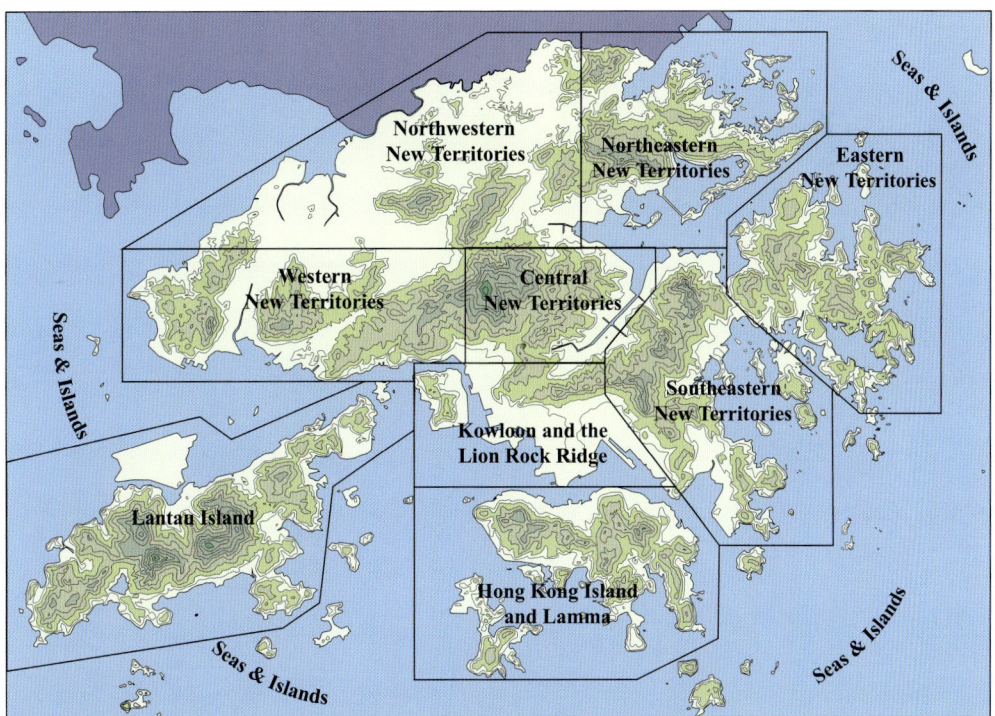

Boundaries of the broad geographical regions recognised in Part Three. Each region is the subject of a separate section.

Acknowledgements:

The authors would like to thank Mr. Felix Poon for making available the marine wildlife photographs on pages: 144, 209, 218, and 226. The GEO/CEDD supplied additional helicopter images (pages: 24, 30, 41, 140, 176, 185, 186, 191, 194, 202, 217, 219, 223, 225, 228, and 229) and the seismic section on page 213. The historical photographs (pages: 65, 117, 174, 180) are reproduced by permission of the Hong Kong Museum of History. Mervyn Peart provided the natural river photograph on page 66. All other photographs are by the authors. We gratefully acknowledge the careful scrutiny of earlier drafts by: Kevin Styles, Tony White, Barbara Owen, Mervyn Peart, and Rod Sewell. We also thank Alberto Dias, Steve Parry, and Janice Shaw for comments and advice. Finally, we would like to thank Colin Day for inviting us to write this book and for providing encouragement during its preparation.

PART ONE
INTRODUCTION

The Beaches of Eastern Sai Kung

MODERN AND ANCIENT ENVIRONMENTS

Preamble

Hong Kong has a largely mountainous terrain, very little flat land, no major rivers, no great forests, and a paucity of mineral wealth. Following the planting of the British flag at Possession Point by Captain Elliot on the 26 January 1841, the British Foreign Secretary Lord Palmerston scathingly remarked that Hong Kong was a "barren rock with hardly a house upon it". In contrast, Hong Kong today is widely perceived as a prosperous cityscape characterised by tall buildings, a dense urban population, and the bustle of Victoria Harbour. Clearly, remarkable changes have occurred since 1841, developments that are advertised worldwide, and are well documented. In contrast, the scenic attractions of Hong Kong beyond the metropolis are little known, and are certainly less well understood.

Prior to urbanisation, the rugged landscape had evolved naturally, with little human interference. Vestiges of the original landscape are still discernible today in the attractive mix of lowland streams, gentle hills, rugged mountains, and the intricately varied coastline. These differing landscapes owe their origins to the kinds of rock that underlie the surface, and to the forces that caused the rocks to bend and fracture. Acting on this solid foundation have been a range of surface processes that include the deposition of sediment by rivers, landsliding on hillsides, and coastal erosion by waves. Human activity has contributed terraced hillsides, dammed streams, reclaimed shorelines, and expanding cityscapes.

This book attempts to both distinguish and explain these varied landscapes, recognising that the present scenery results from the interplay of many factors, including the geology, surface processes, plants, animals, and humans. Employing a broad-based, holistic approach, the authors examine the landscapes in terms of their natural origins, and the distinctive impacts exerted by generations of people. An important aim of this book is to bridge the gulf between academic texts and the less explanatory hiking and photography books.

The Lay of the Land

Hong Kong possesses a variety of landscapes and habitat types. This diversity reflects the climate and Hong Kong's geographical position between tropical and temperate climatic regions. The location

Tai Mo Shan, at 957 m, is the highest mountain in Hong Kong.

results in the climate varying over the seasons. In summer, southerly winds bring hot and humid air to the territory, which results in frequent torrential rainfall. One to two typhoons affect the territory each year. In contrast, northerly winds in the winter cause cool and dry conditions, with surges of cold air from China lowering urban temperatures to just 8–10°C, and rarely as low as 2°C. The height of the ground also influences climate. Altitudes range up to 957 m on the highest peak of Tai Mo Shan, where frosts and the heaviest rains occur. Variability in the weather, in turn, determines the geological processes that shape the land.

Modern and Ancient Environments: The Lay of the Land

The dramatic landscape of Hong Kong has been brought about by the combined influences of: the underlying geology; processes that take place at the earth's surface today; and a variety of human impacts. The oblique aerial photograph above was originally taken from 20,000 feet (6,096 m) on 5 January 1987. Geological influences can be seen in the long straight valley of Sha Tin (centre) and its extension through the Tolo Channel. This alignment reflects a major fault, a line of weakness that has been eroded by surface river processes. A sea level rise since the last Ice Age has resulted in flooding of part of the river valley to form the Tolo Channel. Human impacts are very clear with the extensive urbanisation of Kowloon, Sha Tin (centre left), and Tai Po (upper left).

Aerial photographs are a useful source of geological, geographical, and historical information. The larger oblique photograph is like a model, showing Victoria Harbour in 1987 and its topographical setting. The photograph (right), also taken in 1987, looks more like a map, showing the harbour from directly above (6,096 m) and the streets of Kowloon and Central District. The Lands Department of the Hong Kong government have a territory-wide coverage of aerial photographs of Hong Kong from 1963 to the present day (now in colour), and incomplete coverage dating back to 1924.

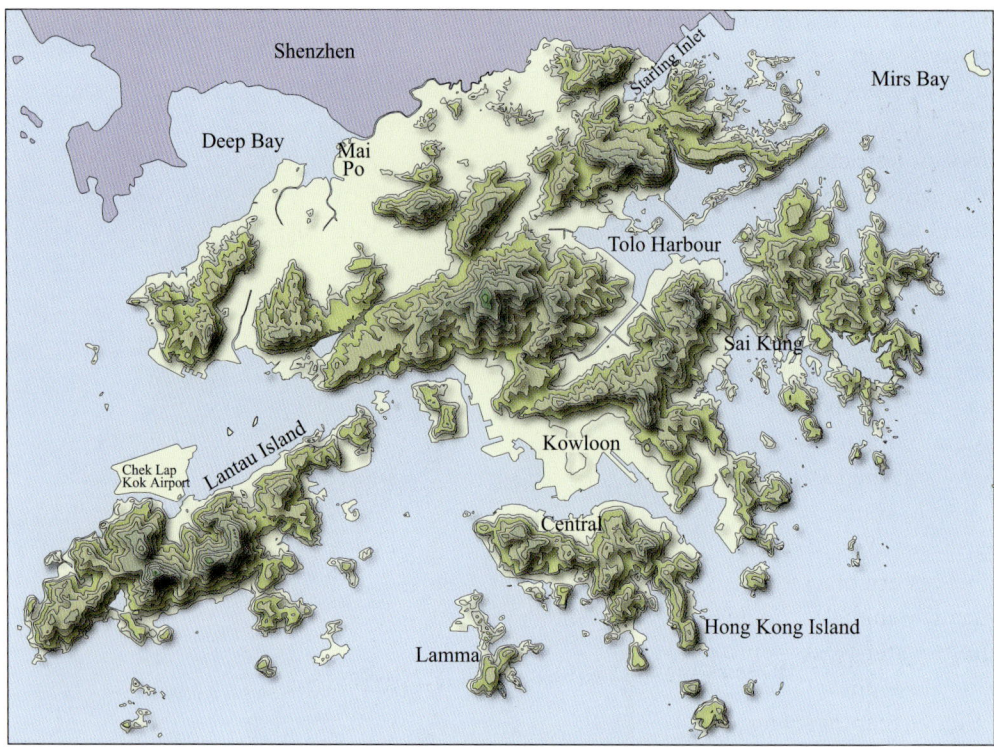

Hong Kong consists mostly of sea, with numerous islands and a highly indented coastline. Mountains rise to over 900 m on Lantau and in the central New Territories. Extensive lowlands are present in Kowloon and the north and northwest.

For example, the hot humid climate causes chemical decay and weakening of the rocks, which, after being waterlogged by heavy summer rain, are then prone to landslides.

The diversity of landscapes also reflects differences between land and sea. Less than half of the territory is land (1,104 km^2, including reclamations); the remainder is covered by the sea (1,800 km^2 in 2006). There are at least 262 islands, the largest of which are Lantau (146 km^2), Hong Kong Island (78 km^2), and Lamma (14 km^2). These islands are actually the summits of hills, the flanks of which were flooded by rising sea levels at the end of the last Ice Age. Similarly, the highly indented coast, in areas such as Tolo Harbour and Sai Kung, was brought about by the flooding of former river valleys.

A variety of environments also occur in offshore areas. The western seas, and Deep Bay, are dominated by the mud-rich, brackish waters of the Pearl River Estuary, whereas Mirs Bay and southeastern Hong Kong waters are part of the South China Sea, with saltier and clearer water. These contrasts in land and sea have generated a considerable range of habitats that provide homes for a multiplicity of flora and fauna.

Hong Kong is situated close to the northern limit of tropical Southeast Asian plants. Despite severe human intervention in the form of clearance for agriculture, fire, and urbanisation, the flora of Hong Kong remains very diverse, comprising at least 2,500 species of land plants. The hilltops and upper slopes where the soil is poor are usually covered with grass, which also occurs in areas frequently affected by fires. Hillsides are commonly clad in scrub. Woodlands are mostly restricted to lower slopes and wetter stream valleys, but are also found as isolated remnants in upland valleys or as Feng Shui groves (IN15, p. 83). Mangrove forests fringe the shores of sheltered inlets such as Deep Bay and Starling Inlet.

Hong Kong also has a great variety of mammals, reptiles, and insects, both in the sea and on land. Marine organisms include corals in the eastern fully-oceanic waters, and pink dolphins (*Sousa chinensis*) in the western brackish areas. Land animals include fresh-water crabs, many kinds of snakes, barking deer, porcupine, pangolin, civet cats, wild boar, and macaque monkeys. Birds are even more varied, with many species making use of the Mai Po wetlands (northwest Hong Kong) to rest and feed during their seasonal migrations.

Reconstructing the Past

The diverse environments of modern Hong Kong merely represent the latest phase in a long history of environmental change on a geological time scale of hundreds of millions of years. Far from being dull and lifeless, rocks should be considered as the pages of a book that reveal the history of the earth. Geology is able to open up an exciting window to the past. Deciphering this ancient history reveals a fascinating story of changing landscapes, shifting climates, and, in certain regions, evolving plants and animals. This is possible because the processes that act at the surface of the earth today are the same processes that have acted in the past: only the scale, frequency, and geographical distribution of the processes vary with time.

This fact was recognised by James Hutton (1726–97), The Father of Geology, who in 1785 enunciated the Principle of Uniformitarianism, which states that "the Present is the Key to the Past". This principle holds that the history of the earth can be explained by examining modern environments. Consequently, by studying processes that operate today and the kinds of sediment that they produce, geologists can understand the origins of ancient rocks comprised of similar materials. Hutton also expounded the "Law of Superposition". This notes that a layer of rock that covers another layer of rock must have been deposited later and is, therefore, younger. Using this simple law, it is possible to establish the sequence of events in an area. Geologists have adopted both principles to reconstruct the way Hong Kong's environment has changed over the last 400 million years. The actual timing of these changes has been determined using radioactive dating techniques (IN01).

IN01 Geological Time and Dating

Several techniques are used to determine when geological events took place. It is possible to date even ancient rocks precisely because naturally occurring radioactive "parent elements" break down into smaller, more stable "daughter elements". This may occur through the spontaneous emission of a variety of particles (e.g. an alpha particle as in the figure) from the nucleus of atoms. These nuclei are comprised of positively charged protons, and neutrons with no electrical charge. The rate at which decay takes place is unique to each radioactive element and is called its *half-life*.

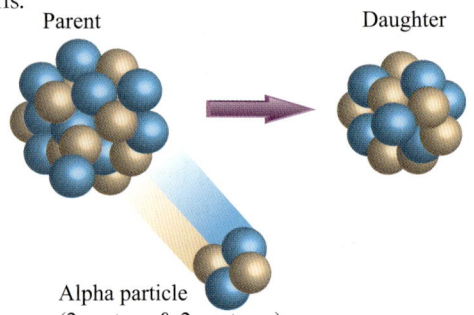

A half-life is the time taken for half of the radioactive parent element to decay to a daughter product. For example, Lead-210 (210 is the total numbers of protons and neutrons in the nucleus) has a half-life of 22 years. This means that an initial 16 g of this element will break down to 8 g after 22 years; and then to 4 g after another 22 years (a total of 44 years), and so on. In practice, it is the ratio of the parent to the daughter element that tells us the number of half-lives. This, in turn, can be used to find the age of the rock in years.

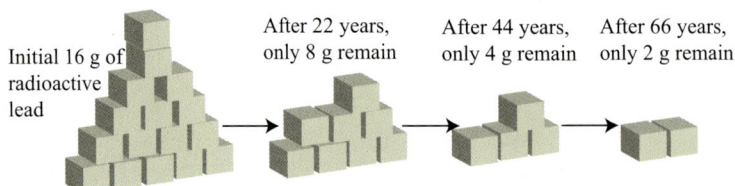

Lead-210 has a half-life of 22 years

The half-life of different elements varies considerably. Uranium-235 takes 704 million years for half of its original material to break down to new daughter products. For Uranium-238, this period is a massive 4.47 billion years. In contrast, Carbon-14 only takes 5,730 years. Different radioactive elements will be chosen for a particular study according to their presence in the minerals that occur within a rock, their breakdown rates, and the time frame of the study.

An Environmental History

The oldest rocks in Hong Kong were deposited as loose sediments 400–360 million years ago. Today, these rocks are only exposed in the northeast, along the northern shores of the Tolo Channel. The sediments originally accumulated near a river mouth that drained mountains to the southeast. Multiple streams carried sand and pebbles across the landscape. Small sand dunes formed locally. Over time, the mountains were eroded and became lower, with finer-grained sand and silt being produced from the mountain remnants. These materials were laid down on a delta floodplain that was periodically invaded by the sea. A similar setting is found today along the lower Nile River in Egypt.

From 400–360 million years ago, a network of rivers flowed northwards over the site of present day Hong Kong. They deposited pebbles and sand in a complex of channels that crossed floodplains close to the coast.

Then, about 360–320 million years ago, the floodplain was inundated by a tropical ocean. Few large rivers entered this sea, so the waters were clear and free of sediment. This allowed creatures with skeletons or shells of calcium carbonate to thrive in the warm, shallow, silt-free waters. Over time, their remains built up a thick layer that was subsequently turned into a rock called limestone. Increasing temperature and pressure, caused by burial under later sediments, converted the limestone into a white, sugary, marble, which today occurs buried below parts of Hong Kong. This rock has been changed so much that fossils can no longer be recognised. Today, we can only speculate about the nature of the life in these seas, but corals, or algae that formed carbonate reefs, are possibilities.

Shallow tropical seas, similar to the photographs, formed between 360–320 million years ago. Coastal swamps developed when the sea level fell, to be replaced by deltas as the sea level again rose.

Subsequently, there was a major fall in sea level, and coastal swamps developed adjacent to a newly exposed land surface. Then, the area was again flooded, with deltas forming on top of the earlier swamps. Something similar to these settings can be seen in the Caribbean Sea with its tropical reefs and mangroves.

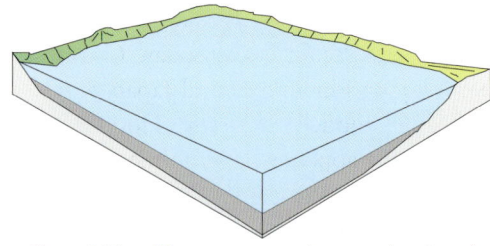

About 210 million years ago, the area that is today Hong Kong was covered by a moderately deep sea that resembled the waters of the present Mediterranean.

The sea level continued to rise, and, by about 280 million years ago, this area lay at the bottom of a deep sea, with silt and mud accumulating on the sea floor. A period of uplift and erosion followed. Then, moderately deep marine conditions were re-established around 210–190 million years ago, with silt and mud again being laid down. Ammonites thrived in these waters. After death, their remains were buried by mud to form fossils that can be found in mudstones dating back to this period. Today, ammonites are extinct, except for one distant relative, Nautilus, that survives in the deep seas of Southeast Asia.

Ammonites are extinct today. They belonged to the same group as the modern Nautilus and squids.

About 165–140 million years ago, the seas were replaced by a chain of volcanoes that stretched hundreds of kilometres along the southeastern coast of modern China. These were fed by magma chambers that lay only 1–2 km below the active volcanoes at that time. Many of the eruptions were extremely violent, throwing out tens of cubic kilometres of ash that fell back to the ground around the volcanic flanks. In some instances, so much material was erupted that the volcanoes collapsed in on themselves to form large circular depressions called calderas (IN40, p. 136). During quieter periods, these calderas may have filled with water.

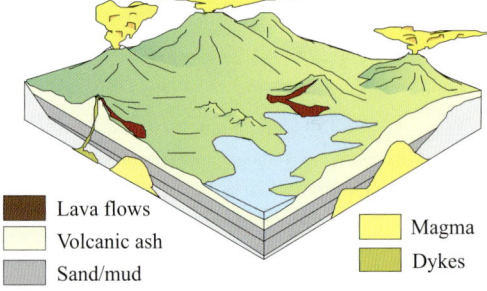

Lava flows
Volcanic ash
Sand/mud
Magma
Dykes

Violent volcanoes, similar to those of the Philippines today, dominated Hong Kong 165–140 million years ago.

Taal Volcano, on the northern island of Luzon, is a modern flooded caldera similar to those that once existed in Hong Kong. The central island is a new volcano growing within the caldera. Another lake fills the central vent of this island.

An Environmental History

These kinds of explosive eruptions occur around the Pacific Ocean today. Countries such as the Philippines and Japan provide excellent models of how Hong Kong might have appeared at this time.

Rainfall was also an important aspect of these ancient landscapes, as it periodically saturated the ash that lay on the ground. This material then flowed downslope to the surrounding lowlands as lahars—dense mixtures of ash and water.

Hong Kong's period of volcanism resembled the modern eruptions of the Philippines. The central vent of Mount Pinatubo (above) is an illustrative example. It is now quiet and filled with water, but it produced the largest explosion of the twentieth century. Vast quantities of ash were thrown tens of kilometres into the sky, before the material settled back onto the flanks of the volcano. Several years later, steam eruptions (far left) were still occurring as rain water came into contact with the hot ash. Much of the ash was also actively reworked in dense and destructive lahars that followed heavy rainfalls (left).

By 140 million years ago, the volcanoes had stopped erupting and the mountains were being eroded away so much that there are no volcanic cones today. The climate of Hong Kong became arid about 100 million years ago and resembled parts of the Middle East. Storms probably caused occasional flash floods. Large angular blocks broken from the mountains accumulated at the base of steep slopes, and fan-shaped lobes of debris formed where intermittent rivers reached lowlands. Pebbles were laid down in winding river channels that were dry for most of the year. Reddened, wind-blown sand dunes developed in some areas.

The desert landscapes of Hong Kong, 100 million years ago, displayed a complex mixture of landforms. Scree slopes (top left) occurred at the base of cliffs. Alluvial fans (bottom left) issued from the foot of mountains as streams dropped their sediments. Ephemeral rivers crossed arid mountainous plains (top right), probably feeding oases. Locally, sand dunes (bottom right) marched across otherwise stony plains.

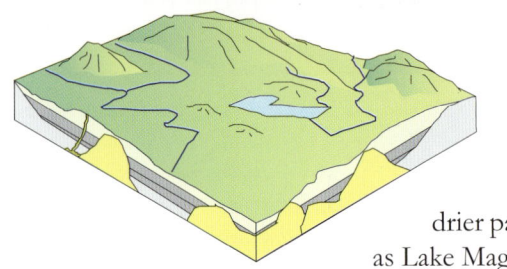

Gradually, the desert environments became a little more moist. By 80–50 million years ago, there was just enough rain to create salty lakes (left). Rivers flowed during the wet season and washed silt and clay into these water bodies. At other times, the lakes dried out and mud settled to the lake floor and cracked. Salt crystals were formed from the evaporating water. Plants that grew during wetter times supported insects that became fossilised in the mud deposits. Modern examples of seasonal lakes like these can be found in the drier parts of the rift valleys of Africa, at places such as Lake Magadi (Kenya), shown on the top left.

Over the last 60 million years, the landscape has continued to evolve. An increasingly hot, wet, and humid climate resulted in chemical decay of the surface rocks. This produced a thick layer of soft, easily eroded material, which was gradually worn away.

Over the last 2 million years, there have been many episodes of rising and falling sea levels that have alternately flooded and exposed the land surface of Hong Kong. These changes are related to the most recent Ice Age, which is only one of several worldwide cold periods that have occurred over a time span of billions of years. Temperatures varied through the Ice Age. During colder phases (glacials), the glaciers in other parts of the world expanded, and the seas fell as water was locked up as ice on the land. During warmer periods (interglacials), the ice melted, the water returned to the oceans, and the seas flooded back over the land. The last major glacial reached its peak about 18,000 years ago. Then, the sea level in Hong Kong fell 120 m below its modern height and the coastline lay 120 km further to the south.

The warmer post-glacial landscape became home to tropical forests by 9,000 years ago. In recent centuries, human impacts have increased rapidly, while natural processes have remained comparatively stable. One consequence, for example, has been the loss of the original rainforests, although there are a few surviving stands of older secondary woods.

The Sai Kung islands represent a flooded landscape created by a rise in sea level at the end of the last glacial period.

PART TWO
LANDSCAPE TYPES AND ORIGINS

The Hunchbacks and Ma On Shan

The Southern Cliffs of Ma On Shan

RUGGED MOUNTAIN LANDSCAPES:
A STORY OF ANCIENT VOLCANIC ERUPTIONS

Long-extinct volcanoes have left an indelible footprint on Hong Kong's landscape. Large parts of the territory were influenced by volcanic activity that produced relatively hard rocks. Today, these commonly form the foundations of the most prominent peaks, including Tai Mo Shan (Big Hat Mountain), which rises to 957 m and is the highest summit in Hong Kong. Other high points underlain by volcanic rocks include: Sunset Peak, Lantau Peak, Fei Ngo Shan, Sharp Peak, and Ma On Shan (p. 12). Although tall cliffs are rare, there are many steep slopes, narrow ridges, and scattered rocky outcrops interspersed with more gentle grassy and wooded areas. Surface boulders are common, resting on relatively thin, stony soils. These rugged mountainous landscapes occur widely in the New Territories, Lantau Island, southern Hong Kong Island, eastern Hong Kong, and throughout most of the Country Parks. The influence of volcanic rocks on the shape of the land surface can also be observed along the more dramatic coastlines, with their steep cliffs and angular, rocky shorelines.

The volcanic rocks were created, not by slow moving lava flows, but by violent eruptions that took place mostly between 165–140 million years ago. Fragments thrown out of these volcanoes originated from hot sticky magma that accumulated

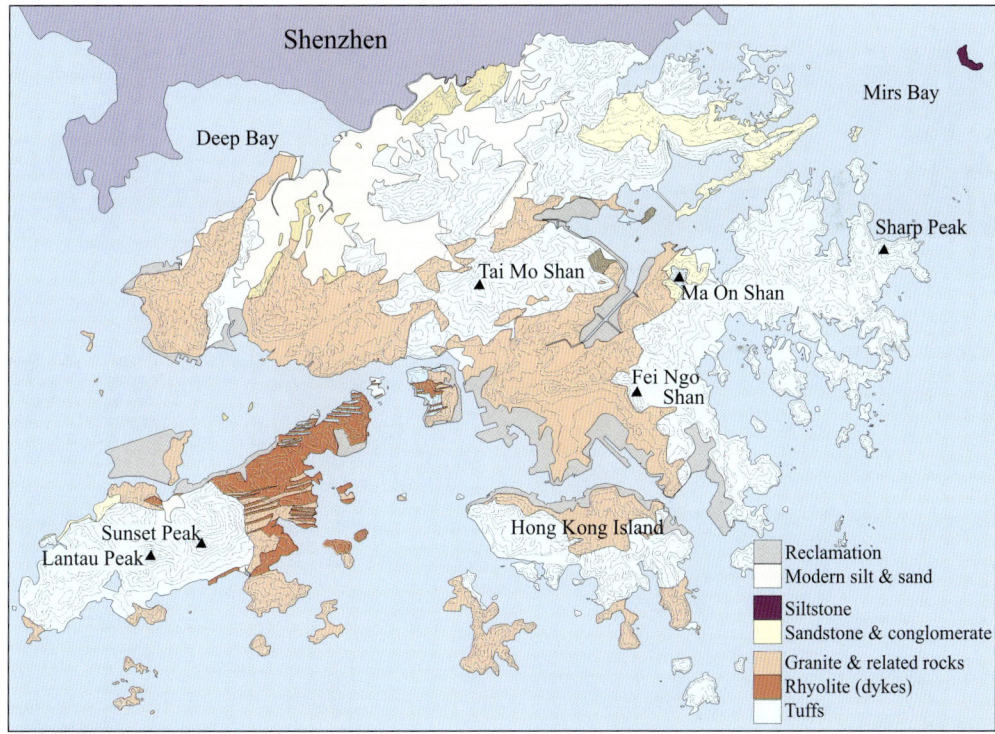

Hong Kong is dominated by volcanic rocks called tuffs (IN02, p. 15) that originated as ash thrown violently out of multiple volcanoes. Today, these igneous rocks are exposed at the surface over about 50% of the territory, mainly occurring in the eastern and northern New Territories, western Lantau, and southern Hong Kong Island.

Volcanic rocks in Hong Kong were originally formed as a result of subduction along a plate boundary. This occurred when one rigid plate was forced below another and melted as it descended into the earth's interior, where it reached areas of higher temperature. Magma was produced, which then rose through the overlying rocks to form magma chambers below volcanoes. Occasionally, this material reached the surface, giving rise to violent eruptions.

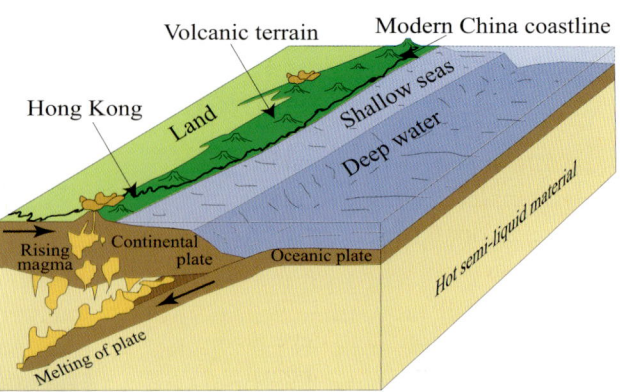

The photograph above shows the northern slopes of Ma On Shan viewed from a northern ridge on Buffalo Hill. The photograph below is of Sharp Peak—the dramatic backdrop to several superb beaches along the eastern coastline of the East Sai Kung Country Park. Neither of these mountains are especially high (705 m and 468 m respectively), but they both illustrate the rugged, angular nature of the terrain that forms on volcanic rocks.

in reservoirs deep below the surface. These volcanoes were part of an extensive volcanic terrain along, what is today, the south China coastline. The eruptions took place along a linear plate boundary (figure above). In this region, the southern plate was driven northwards, melting as it descended deeper below the northern plate, producing magma.

Similar plate boundaries occur today around the edges of the Pacific Ocean and are responsible for the active volcanoes in places such as the Philippines, Japan, and New Zealand. The rate of supply of the magma controls the eruption frequency. Magmas produced in these settings are rich in silica, which makes them very viscous (sticky). Consequently, they tend to resist movement (flowing), so that eruptions are particularly violent. Lava flows are infrequent and of small volume, travelling only short distances. Another factor in the explosiveness of these eruptions is the expansion of dissolved gas, which is released as the magma rises through the crust and ambient pressures are reduced. Eruptions of this kind generate large quantities of pulverised rock and clouds of ash.

In Hong Kong, the bulk of the erupted material formed fiery clouds of volcanic ash (about 85%), with rhyolite lavas (IN02, p. 15) being less common. The ash was

IN02 Extrusive Igneous Rocks

Igneous rocks originate from the cooling and solidification of hot liquid magma, which is a complex mixture of many elements, dominated by silicon and oxygen. The rocks that form are very varied in their appearance and primarily depend on the chemistry of the original magma and the geological environment in which they cool.

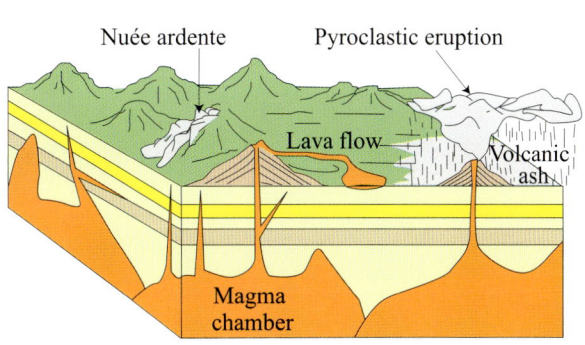

Two major types are recognised. Extrusive rocks are formed at the surface, whereas intrusive rocks remain within the earth's crust (IN04, p. 22).

Hong Kong's extrusive rocks are dominated by tuffs. Lava flows are relatively rare. Tuffs form from layers of ash that settled to the ground surface after being erupted violently into the atmosphere. In some cases, ash mixed with gases from an erupting volcano and the atmosphere to produce a hot dense glowing cloud (a nuée ardente) that flowed down the flanks of the volcano, hugging the ground. The resulting deposit is a particular type of tuff called an ignimbrite.

Extrusive Rocks in Hong Kong:

Crystal tuffs are dominated by mineral fragments. Larger crystals may occur, surrounded by smaller minerals that are too small for the eye to see.

Ignimbrites contain flat volcanic fragments that lie parallel to each other. These were formed when the ash was still hot, loose, and squeezed under its own weight.

A large proportion of the tuffs in Hong Kong consist of a mixture of small mineral and larger, angular rock fragments. Most are a speckled grey colour.

Modern weathering and rock decay pick out larger volcanic fragments on rock surfaces. This photograph shows a volcanic bomb that was thrown out of a volcano and which settled into finer-grained ash.

Rhyolite lava contains minerals with a high silica content. This example from Sai Kung is dark grey and fine-grained, with minerals that are too small to see, except for some larger scattered feldspar crystals.

mostly fine-grained, consisting of angular crystals, volcanic glass, and rock fragments ripped from the sides of the vent. Two main styles of eruption occurred. Some of the material was thrown out as vast, billowing, fiery clouds (nuées ardentes, IN02, p. 15) that flowed at high velocity down the flanks of the volcanoes. Alternatively, air-fall eruptions occurred, during which ash was blown out vertically, up to hundreds or even thousands of metres into the atmosphere, before falling back to the ground. After subsequent events buried the ash, increasing temperature, pressure, and fluids within the rocks caused the loose particles to change into solid tuff. It is this rock that dominates Hong Kong's volcanic landscapes and rugged mountains.

Variations in the size of particles within tuffs produce contrasting landscape types. For example, the most angular and

The Stone Trail, in eastern Sai Kung, follows a stream that drops over several large waterfalls. The photograph shows one of the smaller falls, about 7 m high.

Tuffs are formed from loose volcanic ash that has been turned into solid rock. Tuff is hard and resistant to erosion, forming steep slopes and narrow gorges such as along the Stone Trail (above) in eastern Sai Kung Country Park.

Rugged Mountain Landscapes

These large rounded boulders, along a Lantau Island footpath, were formed by weathering of solid rock and removal of the fine-grained decayed material (mainly clay). This particular boulder field may also have been involved in a landslip. Boulder fields on hillsides are a common feature of areas underlain by coarser-grained volcanic rocks.

craggy areas are generally developed on fine-grained tuffs, which are more resistant to both chemical and physical decay (IN51, p. 177), and to erosion. In some places, rivers cut through these rocks to form narrow gorges. An excellent example occurs along the Kap Man Hang, a stream in eastern Sai Kung Country Park. The Stone Trail (photographs, p. 16), a difficult walking route, follows part of this river as it passes through several tight gorges and descends a series of spectacular waterfalls. In contrast, coarse-grained tuffs are more easily weathered and tend to produce more rounded hills. The weathered layer at the surface is thicker, and boulder-fields (IN10, p. 39) are strewn across the ground, resulting in an appearance similar to that of granite landscapes (p. 21).

In parts of Hong Kong, such as the Pat Sin Leng in Plover Cove Country Park, volcanic and sedimentary rocks occur together. Both have been tilted by past earth movements, giving rise to steep, rocky inclines on one side of the mountain

The steep southern scarp face of the Pat Sin Leng is shown here cutting across inclined layers of volcanic and sedimentary rocks. The highest point in the photograph is Wong Leng, which rises to 639 m above sea level.

IN03 Escarpments and Cuestas

Inclined layers of volcanic or sedimentary rocks tend to produce distinctive landforms. Ping Fung Shan provides a good example, with a thin layer of conglomerates and sandstones overlying a thick sequence of tuff. These layers dip to the north and form the gentle slopes that descend to Sha Tau Kok Hoi. The steep southern slopes cut across these rock layers and form a craggy scarp slope.

Rugged Mountain Landscapes

These steep slopes, above the small beach at Long Ke Tsai (southeastern Hong Kong), are underlain by volcanic rocks. At the coast, erosion by the sea has produced a laterally extensive fringe of rocky outcrops up to about 10 m high and which locally form steep cliffs.

and gentle slopes on the opposing side that follow dipping rock layers. These asymmetrical ridges are called cuestas (IN03, p. 18).

Volcanic rocks also form many of the coasts of Hong Kong. Where they are exposed to the South China Sea, they tend to form prominent vertical cliffs. In these settings, powerful ocean waves undercut the tuffs, which then collapse. However, because the rocks are hard, they remain upstanding, except where they have been weakened by faults (IN08, p. 35) or joints (IN09, p. 37). The best examples of these landscapes can be seen along the southern and eastern shorelines of Hong Kong.

The rugged mountains of Hong Kong constitute its most dramatic natural settings. However, there are other landform types that also demonstrate the relationship between scenery and rock type, including rounded hilly landscapes (underlain by granitic rocks), as noted in the next section.

The cliffs of Bluff Island, in southeast Hong Kong, were formed by high energy waves that eroded the tuffs along the exposed coastline. Where faults or joints (cracks) occur, the rocks are weaker and more easily eroded, and sea caves develop.

19

Rounded Hilly Landscapes

Granite Boulders of Lung Kwu Tan, Western New Territories

ROUNDED HILLY LANDSCAPES:
THE ROOTS OF VOLCANOES

Today, the roots of ancient volcanoes create very distinctive landscapes in Hong Kong. These roots were the magma chambers that were once located 1–2 km below the ancient land surface. Over time, the hot molten magma cooled and solidified to form the rock granite. Millions of years of subsequent erosion stripped away the overlying rocks (IN04, p. 22), exposing the granite over one-third of the land area of Hong Kong, forming bold, rounded rock surfaces.

These granite regions tend to be lower and more gentle in appearance than the volcanic landscapes described in the previous section. Rounded boulders are generally strewn across the surface (p. 20). In some areas, there are dense networks of deep gullies cut into the upper layer of chemically decayed (weathered) rock.

The granitic rocks that underlie these areas extend across much of the western New Territories, including the Tai Lam Country Park and the Castle Peak Range. They also make up the Lion Rock Ridge, the Chi Ma Wan Peninsula, and parts of the southern offshore islands. Granite also forms the flat ground and isolated hills below urban Kowloon and Central.

The hot, humid, sub-tropical summers in Hong Kong cause chemical breakdown of granites. Consequently, granite terrains are usually characterised

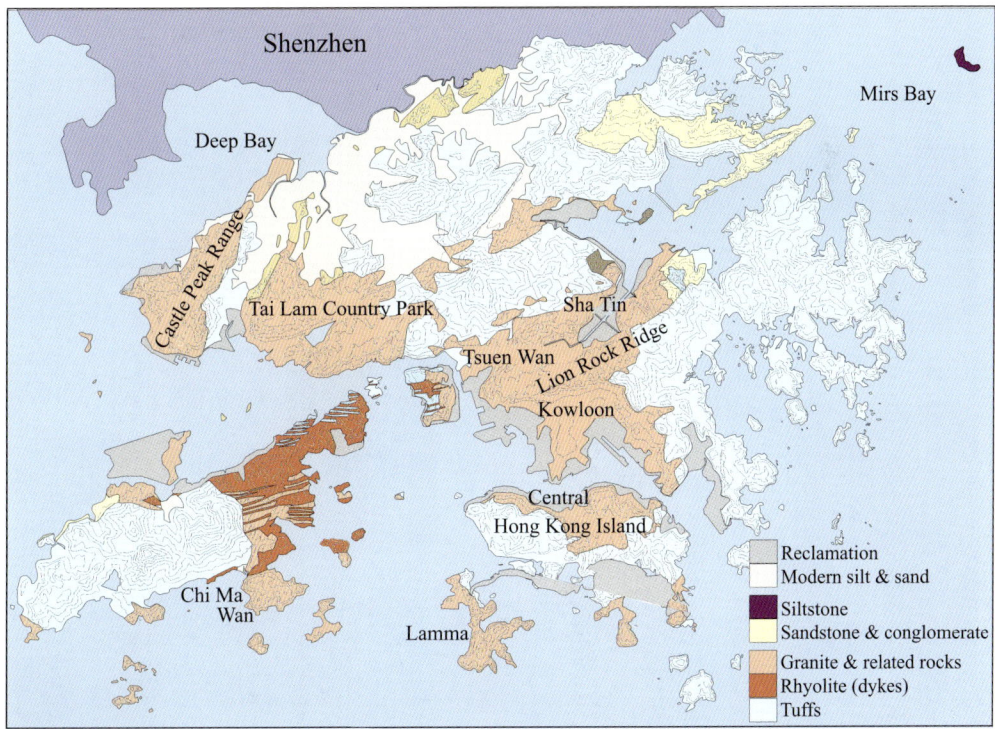

Granitic rocks cover about 35% of the land area of Hong Kong, mainly forming gently rounded hills, although they also underlie the relatively flat areas of Kowloon and the steep slopes of northern Hong Kong Island.

IN04 Intrusive Igneous Rocks

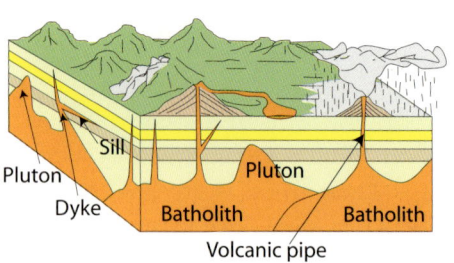

Igneous rocks solidify from magma that has cut its way into the crust or which has erupted at the surface. Rocks formed in the former manner are referred to as intrusive. They may have cooled within small-scale, sheet-like dykes (cutting across the rock layers) or sills (lying parallel to the rock layers). Larger bodies of magma form plutons, and the largest accumulate in batholiths. Most magma in Hong Kong originated in plutons. The mode of intrusion is important to the appearance of the rocks. Small intrusions cool relatively quickly, and crystals only grow to a small size. Larger intrusions cool very slowly, and crystals can grow larger. The chemistry of the magma also controls the final rock type. Magma that developed in Hong Kong originated from melting along a plate boundary (p. 14; IN23, p. 105); it is rich in silica. Upon cooling, this magma type generates granite (in plutons) and rhyolite (in dykes and sills).

Intrusive Rocks in Hong Kong:

Granitic rocks form through slow cooling of magma in major intrusions. Crystals can grow very large. The minerals are dominated by several types of feldspar (white/grey to pink), less commonly by quartz (glassy, grey), and by a small percentage of dark minerals, usually hornblende or biotite. Variation in the proportions of these minerals is the basis for recognising different kinds of granitic rock. Granite (left) contains mainly alkali feldspars that are rich in sodium and potassium. Granodiorite (top right) has a little more calcium-rich feldspar (plagioclase), whereas monzonite (lower right) has less quartz and roughly equal proportions of alkali and plagioclase feldspars.

Basalt is a dark grey-to-black, very fine-grained rock. It is rare in Hong Kong, occurring in dykes, as in this photograph. In other parts of the world, basalt is much more common, forming mainly as lava flows at the surface.

Rhyolite forms as a lava or within dykes. In Hong Kong, it is mainly present in dykes. When weathered, rhyolite may develop parallel striations on the surface (top left) caused by minerals that lined up parallel to one another as the magma was intruded. In some cases, large feldspar crystals have grown within a finer-grained matrix (centre and lower right).

Rounded Hilly Landscapes

The rounded hills, strewn with granite boulders behind Lung Kwu Tan (western New Territories), are typical of granite landscapes.

by thick sequences of loose, chemically decayed rock, called a weathering profile (IN51, p. 177). In some cases, the layer of decayed rocks can be well over 100 m thick. These loose and weak materials are particularly susceptible to erosion and gullying, especially after vegetation has been removed. Deforestation has occurred widely in Hong Kong in the past, and this has resulted in severe erosion of many granitic areas, although reforestation efforts since World War II have reduced the problem and hidden many of the old scars on the landscape.

Boulders are commonly strewn across granite hills. They are unaltered, rounded rocks that formed as corestones (IN10, p. 39), surviving the weathering process and remaining on the land surface as the surrounding, decayed material was removed by erosion. The boulders form a distinctive element of the granitic hills in Hong Kong: piles of rock called tors.

Gully erosion can be very deep, as can be seen along this old military road in the Castle Peak Range.

The Castle Peak Range, in western Hong Kong, shows dramatic erosion, particularly along ridge tops. Many other granitic areas in Hong Kong experienced similar erosion, but extensive reforestation since World War II has been gradually hiding the scars on the landscape.

Granite tors are common in the western New Territories.

Ridges and Colourful Landscapes

*Ap Chau,
Northeastern New Territories*

RIDGES AND COLOURFUL LANDSCAPES: ANCIENT SEAS, RIVERS, LAKES, AND DESERTS

Over the last 400 million years, environments in Hong Kong have changed radically (p. 7–10). There have been several phases of volcanism, but at other times, ancient seas, river plains, lakes, and deserts have developed. Sediments that accumulated in these varied settings were subsequently buried and then transformed into rocks. Today, these layered rocks form beds (IN05, p. 26) that later earth movements have tilted to varying degrees. In some cases, the layers are gently sloping and form asymmetrical ridges, or cuestas (IN03, p. 18). In other instances, the beds have been upturned so that they are almost vertical. The vertical beds, in turn, produce distinctive linear ridges. These sedimentary rocks are confined to the mainland and islands of northeast Hong Kong (p. 24), and small parts of northern Lantau. They underlie about 5% of the total land area of Hong Kong.

The Bluff Head Ridge to the south of Plover Cove (above) is underlain by very steeply dipping sandstones and mudstones that were formed in river systems about 400–360 million years ago.

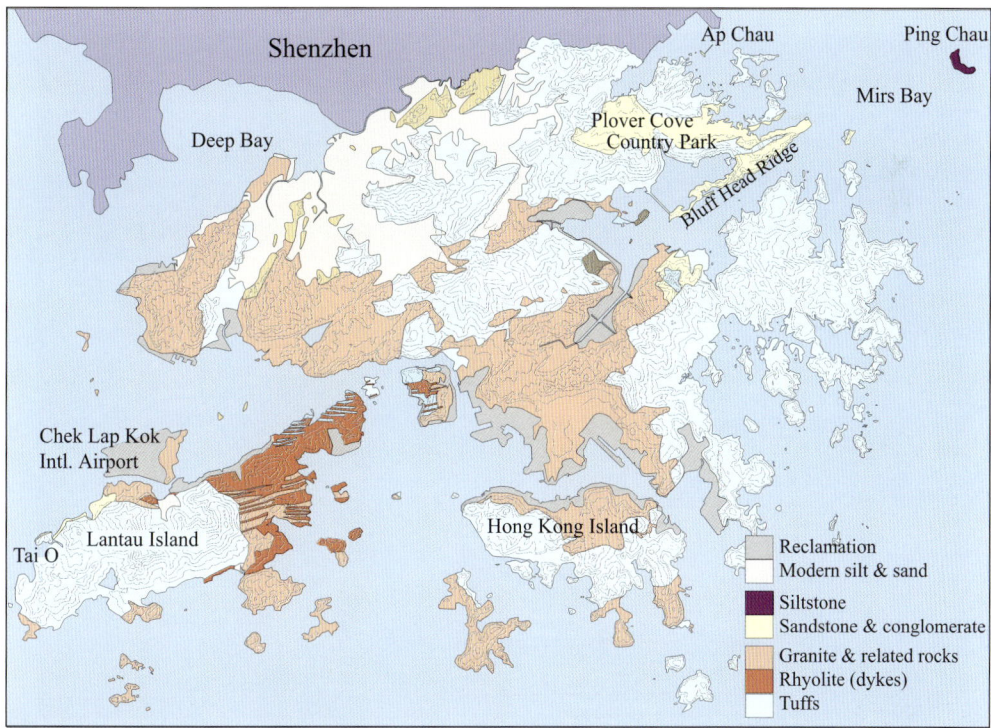

Sedimentary rocks in Hong Kong mainly occur in northeastern Plover Cove Country Park and on the island of Ping Chau. Sedimentary rocks are also present along the northern coast of Lantau, between Tai O and Chek Lap Kok.

IN05 Beds, Bedding Planes, and Time

Sedimentary rocks originate as a series of layers or beds. Each bed represents a single depositional event. The laying down of the sediment may have been rapid, taking only a few seconds to occur, or the bed might represent the slow accumulation of fine-grained clay and silt over a long uninterrupted period.

The beds are separated by discontinuities called bedding planes. These are breaks along which the rocks tend to split relatively easily. They usually indicate a period when no sediment was being deposited. There is no easy way to determine how long that time period was. It could range from several seconds to many years. Geologists believe that most of geological time

Thin, steeply dipping, mudstone beds control the shape of this hillside on Shek Uk Shan, Sai Kung Country Park.

is represented by these gaps in the record. Beds may also vary substantially in thickness from a few millimetres to several metres. Beds can produce a stepped-appearance as the ground follows the rock layers. The thickness of beds controls the size of these steps and the general appearance of the landscape. Note the contrast between the thin, steeply dipping (sloping) beds in the upper right photograph with the scenery in an area of thickly bedded and gently dipping sedimentary rocks (below).

The sandstone beds above (Ping Fung Shan, Plover Cove Country Park) are thicker than the sedimentary layers in the top photograph. They are also more gently inclined, imparting a different texture to the two landscapes.

Grain Size (mm)	Name of the Size Class	Sediment Name			Rock Name		Example
>256	Boulders	Boulders, cobbles, pebbles, granules			Conglomerate if round in shape, or breccia if the particles are angular		
64–256	Cobbles						
4–64	Pebbles						
2–4	Granules						
1–2	Very coarse sand	Sand			Sandstone		
½–1	Coarse sand						
¼–½	Medium sand						
⅛–¼	Fine sand						
1/16–⅛	Very fine sand						
1/16–1/256	Silt	Silt	Mud (a mixture of silt & clay)		Siltstone	Mudstone	
<1/256	Clay	Clay			Claystone		

Classification of detrital sediments and sedimentary rocks. Both systems use grain size. Loose pebble and sand deposits are shown on the upper and centre right. The rock siltstone (lower right; with white calcite) is composed of finer particles.

Sediments can be divided into chemical, biochemical, and detrital types. After they are laid down, processes associated with burial will turn the loose materials into solid rocks. Chemical rocks initially precipitate from a solution; they are rare in Hong Kong. Examples occur on Ping Chau (northeastern Hong Kong), where salts were laid down, together with mud, in a desert lake that periodically dried out. Biochemical rocks, such as limestone, originate from an accumulation of the skeletons of organisms. Coal is another example because the carbon originates from plants that extracted carbon dioxide from the atmosphere.

Sedimentary rocks and sediments that occur in Hong Kong are mostly detrital in origin and are distinguished by the size of their constituent particles (table above). Detrital sediments result when older rocks are eroded into fragments. These are then washed down rivers, blown by the wind, transported by glaciers, or moved in other ways. The particles eventually come to rest in rivers, lakes, seas, or other settings.

Shell accumulations often develop on beaches and when partially consolidated are called coquinas (top right). The solid rock equivalent is limestone (second right; small arrows show individual fossil shells). Limestone is found rarely in Hong Kong. However, when this rock is subjected to high pressure and temperature, it changes to marble (centre), with all evidence of the original fossils being lost. This latter rock occurs in several parts of Hong Kong but only underground, below Yuen Long, Ma On Shan, and Tung Chung. Coal (right) once occurred but long ago was changed by pressure and heat (due to deep burial) to graphite schist (bottom right), which occurs on West Brother Island, north of Lantau.

Ridges and Colourful Landscapes

The rocks along the northern side of Tolo Channel are made up of sedimentary sandstones and mudstones, laid down in river channels and on floodplains around 350 million years ago. Originally, they were horizontal, but earth movements have altered their orientation so that they now stand near-vertically. The photograph shows alternating sandstones (light colours) and mudstones (dark) at Bluff Head.

Colours in sedimentary rocks vary considerably. The upper photograph shows dark grey shales interbedded with red- and orange-coloured siltstones caused by iron reacting with oxygen in the atmosphere. The lower left photograph shows red oxidised iron that has accumulated along cracks (joints). In contrast, the black colours in the lower right photograph are caused by manganese oxides.

The layering of sedimentary rocks exerts a pronounced control over the shapes of hills and islands (IN03, p. 18), but these rocks are also characterised by a wide variety of colours that add to the beauty of their landscapes. Iron is particularly important in determining colour. Depending on its precise chemical combinations, it may be red, brown, yellow, green, grey, or even black. Only small quantities of iron are required to generate strong colours. For example, many of the sandstones in the Pat Sin Leng and Port Island Formations (IN06, p. 29), in northeastern Hong Kong, were originally laid down in rivers that periodically dried out. These have deep red colours due to iron combining with oxygen in the atmosphere. Associated finer-grained mudstones and siltstones tend to have purple colours, which also developed at a time when there was a dry landscape (100–80 million years ago). Red colours can also develop in other environments. For example, iron-rich minerals decay near the surface during weathering, releasing iron into water. Then, the water flows along cracks (joints) and, if oxygen is present, iron oxides, such as haematite (Fe_2O_3) and limonite ($FeO.nH_2O$), may be deposited along the joints, as illustrated in the lower left photograph.

Other elements may also add distinctive hues and variety to landscapes. Manganese, for example, tends to produce black, brown, and purple colours. Carbon usually results in black. Copper tends to yield a range of green colours.

Geologists often group rocks together into formations (IN06, p. 29) based on a variety of evidence. The grain size, composition, colours, and fossils, allow geologists to work out the original settings in which they were laid down. They do

IN06 Sedimentary Formations

Geologists will often group rocks into formations that display similar characteristics and which can be mapped across an area of interest. A formation may range from less than 1 m to several thousand metres in thickness. Sedimentary formations can be distinguished by their grain size, structures, or any other criteria that is useful. Each formation is given a name, usually based on the location where its properties are best exemplified and the rocks are best exposed. The sedimentary rocks in Hong Kong have been divided into eleven formations. The adjacent figure shows the distribution of land and sea when each of these formations were deposited. The modern China coastline and Hong Kong are shown for reference.

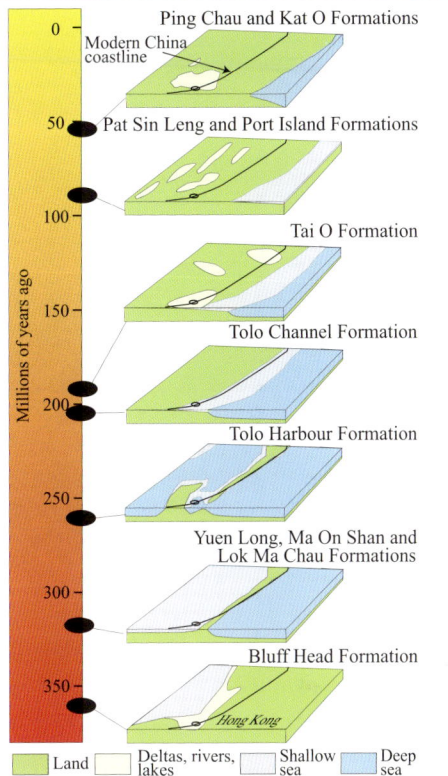

Some of the sedimentary rocks formed on the land, and others below the sea. The lower diagram shows the environment in which each formation developed. The Bluff Head Formation (400–360 million years old) (conglomerates, sandstones, and mudstones) was laid down in deltas and rivers near the sea. The Yuen Long and Ma On Shan Formations (360–320 million years old) (marble) were formed in shallow seas. The Lok Ma Chau Formation (organic-rich mudstones) accumulated in swampy deltas. Mudstones of the Tolo Harbour Formation (290–250 million years old) were laid down in shallow water. The Tolo Channel Formation (210–190 million years old) mudstones were deposited in relatively deep water (>20 m), whereas the Tai O Formation sandstones and siltstones were laid down on a river plain. The Pat Sin Leng and Port Island Formations (about 100–80 million years ago) (conglomerates and sandstones) were formed in rivers. The Kat O (breccias) and Ping Chau Formations (within the period 80–50 million years ago) (mudstones and siltstones) were deposited in arid locations. The former developed when large blocks of rock accumulated at the base of a cliff. The Ping Chau Formation originated in a very shallow, ephemeral, salty lake.

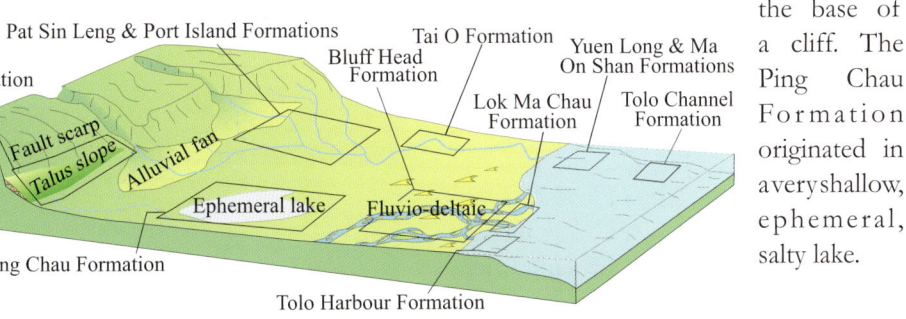

Differential erosion operates at a variety of scales. The left photograph was taken on Ma Shi Chau in Tolo Harbour. A series of beds occur that dip (slope) to the right. The harder siltstones stand about 5–10 cm above the softer mudstones, producing a series of small parallel ridges. On a larger scale, the right photograph shows a prominent cliff and ridge formed by hard conglomerates, also dipping to the right, in western Sai Kung Country Park. Note the band of conglomerates (table, p. 27) continuing into the background and controlling the form of the hillside. The softer rocks here are volcanic in origin.

this by comparing the ancient rocks with similar deposits accumulating in analogous modern environments. This process of comparison has been carried out for the rock formations in Hong Kong. The results indicate that there have been many changes in the prevailing environments over the last 400 million years (p. 7–10).

Landscapes in areas underlain by sedimentary and other layered rocks exhibit distinctive topographical features caused by differential erosion (IN07, p. 31). The combined effects of dipping sedimentary beds (IN05, p. 26) and the occurrence of alternating hard and soft layers produce ridges on a variety of scales, ranging from the Ping Fung Shan Range (IN03, p. 18), at over 600 m, to smaller irregularities, such as those on the islands of Ping Chau and Chek Chau in Mirs Bay (adjacent photographs).

Differential erosion affects the various types of rocks in other ways. They may be fractured by cracks (joints), or by faults. Differential erosion commonly lowers the land surface along these lines of weakness, which is the subject of the following chapter.

Chek Chau lies at the entrance to Tolo Channel and consists of dipping (sloping) volcanic rocks overlain by red sedimentary sandstones (above). The latter originally formed on a dry plain crossed by ephemeral rivers.

Ping Chau consists of gently dipping fine-grained mudstones and siltstones. These form thin layers that control the shape of the island. Note (above) the step-like pattern of the wave-cut platform. The overall shape of the island follows these sloping rocks.

IN07 Differential Erosion

The word rock is usually synonymous with "hard". However, some rocks are less hard than others. Sedimentary rocks, such as mudstone, shale, and siltstone, are relatively soft, whereas sandstone or conglomerate are generally harder. Natural weathering and erosion processes tend to emphasise these differences. This differential erosion is a

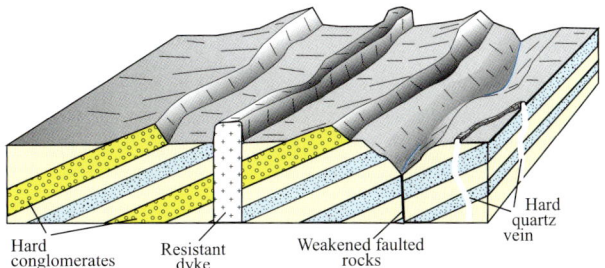

The diagram shows a series of inclined sedimentary rocks. Some layers, such as the conglomerates, tend to be harder than others and stand proud at the surface. Other rocks, such as the dyke and the quartz vein, are also resistant to erosion and form upstanding elements in the landscape. In some cases, rocks can be weakened by faulting, in which case they are preferentially removed by erosion to produce valleys.

powerful force in landscape evolution. Consequently, resistant layers will eventually stand proud above a landscape, whereas softer rocks will form lower areas. The effect occurs at all scales from a small outcrop to the highest mountains. Commonly, rocks that form mountain peaks are harder than the surrounding foothills, those that make ridges are harder than the rocks that comprise valleys, and any upstanding outcrop is harder than adjacent depressions.

Harder rock layers are sometimes caused by secondary effects. Here a prominent siltstone ridge stands about a metre above its surroundings. This layer has been made resistant by a network of quartz (white) veins, as shown in the inset.

Mai Po Mangroves, Northwestern New Territories

LOWLANDS AND VALLEYS: FRACTURED ROCKS AND RIVERS

Faults, large and small, are zones of weakness in the earth's crust that control the locations of Hong Kong's rivers and plains. Most valleys reflect the combined influence of fracturing of rocks by faulting and erosion by rivers. This is apparent when the pattern of faults is overlain on a topographical map (figure below).

There are three main fault orientations in Hong Kong. The largest valleys follow a NE-SW alignment, as shown by the Sha Tin Valley and its extension along Tolo Channel (a flooded river valley). Other faults that run parallel to this trend control the configuration of several valleys on Lantau (Pui O to Yam O Wan), the Lam Tsuen Valley, and the Sha Tau Kok Inlet.

Similar NE-SW faults extend from the border to Tuen Mun, and have contributed to the formation of the lowlands near Yuen Long and the mangrove-fringed coastal plains of Mai Po (p. 32). These latter lowlands are underlain by marble (IN11, p. 51) and also owe their existence, in part, to a faster lowering of the surface of these rocks (IN07, p. 31).

Other valleys (e.g. Shek Kong; and below the High Island Reservoir; along the Lei Yue Mun Gap; and the East Lamma Channel follow a NW-SE fault direction. A third fault trend, N-S, guides the orientation of, for example, Long Harbour and the sea floor topography in southeastern Hong Kong.

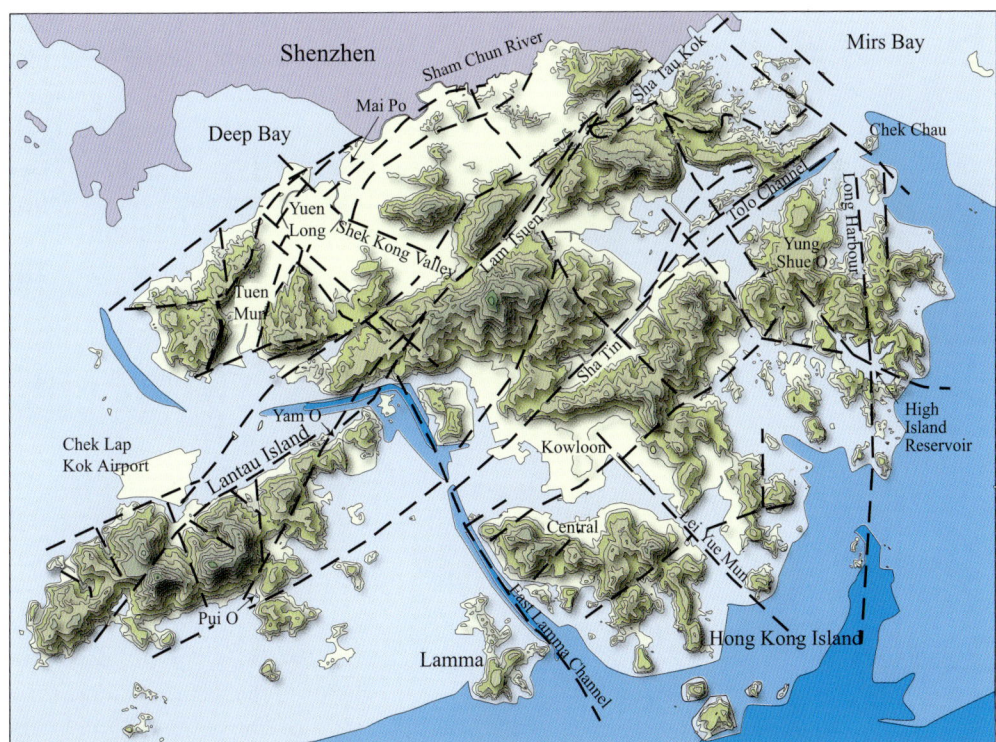

Faults control the orientation of valleys, plains, and sea floor channels in Hong Kong. This can be seen in both the onshore and offshore topography, which displays three major orientations: NE-SW, NW-SE, and N-S.

Lowlands and Valleys

Shek Kong lies at the upper end of a fault-controlled valley that extends through Kam Tin to the Yuen Long Plain in the left background. The peaks of Kai Kung Leng, Tai To Yan, and Kwun Yam Shan are dominated by volcanic rocks.

The inset shows white clay (kaolin) along a fault (IN08). The clay formed as a result of decay of feldspars in the crushed rock.

Movement along faults causes rocks to fracture or even melt due to frictional heating. Faults may occur as sharp breaks, form wide zones of crushed rock (IN08, p. 35), or occur as sets of closely spaced fractures. They promote an increase in the depth of weathering because they create planes along which water can flow and enhance chemical decay (left photograph). Faults may be small-scale, just a few metres long and with a displacement of only a few centimetres, or they may be much larger extending over tens of kilometres and with much larger movements. Most faults are steeply inclined and form straight valleys, such as occur at Shek Kong and Kam Tin (photograph above).

Hong Kong faults form part of a much more extensive structural trend. This is referred to as the Lianhuashan Fault Zone, which extends along the coast of southeast China and controls the regional orientation of many of its landforms.

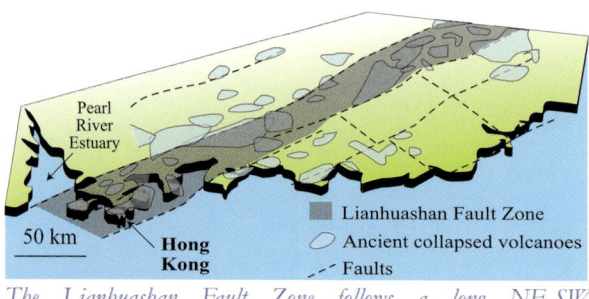

The Lianhuashan Fault Zone follows a long NE-SW trend across much of southern China and Hong Kong.

Several types of faults can be distinguished, based on their style of movement (IN08). As noted earlier, the dominant faults in Hong Kong follow a NE-SW

IN08 Types of Fault

Faults are fractures in rocks caused by movements of the earth's crust. They also involve movement of the rocks on one or the other side of the fault plane. Several types can be distinguished, depending on the direction of movement. A normal fault occurs when the rock above a fault plane moves downwards. In contrast, reverse faults display an upwards movement of the rocks above the fault plane. A thrust fault is similar to a reverse fault, except that the fault plane lies at a more gentle angle. Strike-slip faults involve only lateral movement. In some situations, both vertical and horizontal movements can take place and the term "oblique fault" is applied.

The type of fault that develops depends on the stresses within the rocks. Normal faults occur when rocks are pulled apart (tension), whereas reverse and thrust faults form when rocks are squeezed (compression). Strike-slip faults occur where stresses are mainly lateral.

Faults may be sharp breaks (right photograph), or they may be associated with wide zones of crushed rocks (lower photograph). Sometimes they are filled by mineral veins formed at a later stage.

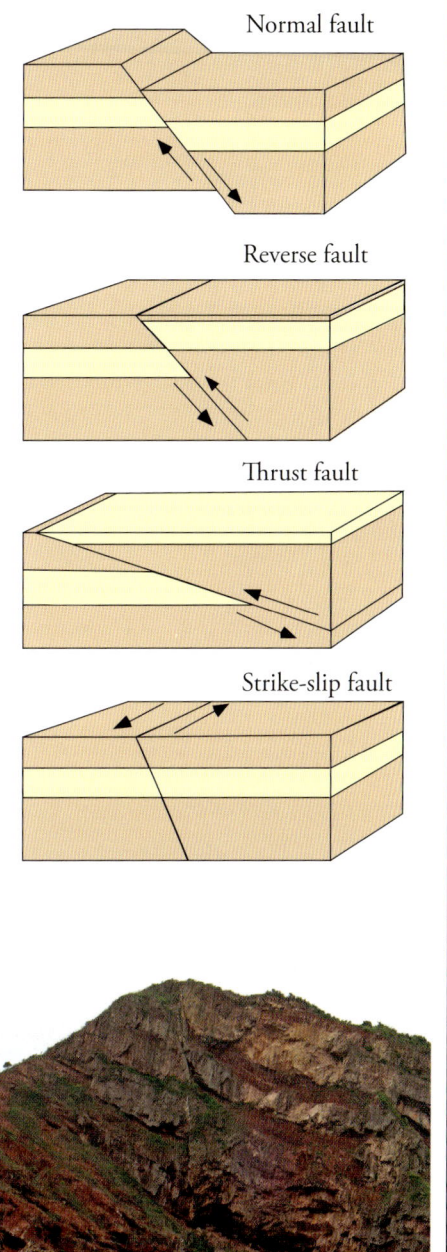

A crush zone (between the lines) marks a fault. The offset is an illusion caused by a flat platform.

The photograph above shows a vertically oriented fault immediately below the summit of a small hill on Chek Chau (northeastern Hong Kong). Note the sharp break with a distinct offset in the dipping sandstones on either side of the fault.

orientation and control the alignment of the major valleys. These particular faults are mainly of the strike-slip type (IN08, p. 35) and are spaced about 6–12 km apart. Movement along the fault planes has been predominantly sideways (horizontal) and has occurred intermittently through the last 300 million years or so. Total movements, up to 3 km, have been responsible for changing the relative position of rocks on either side of the faults and for determining the shape of the ground. Lantau Island provides a good example. It is crossed by two major strike-slip faults that have caused displacement of the rocks (A and B in the adjacent figure) and have elongated the shape of the island.

Faults are not the only kinds of fractures that affect landscapes. Joints (IN09, p. 37) are also cracks, but they do not involve movement on either side of the fracture surface. They tend to produce

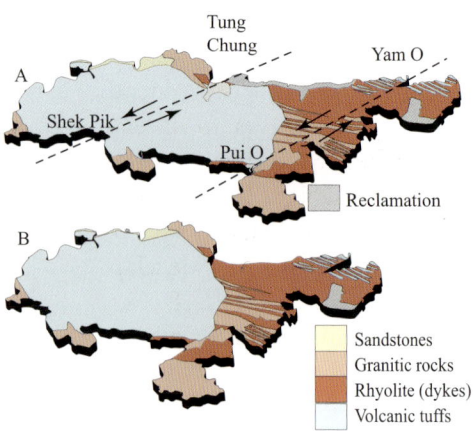

Strike-slip faulting has changed the shape of the land that is now Lantau Island. Repeated slow movements along two NE-SW-trending faults have shifted the rocks sideways by up to 3 km (Figure A). If the effects of these movements are removed, the original relationships of the rocks can be reconstructed. This has been done in Figure B. Note the resulting circular shape for the outcrop of the tuffs on western Lantau, which reflects their accumulation within an ancient caldera (a circular depression created by the collapse of a volcano).

This photograph shows a series of well-developed, parallel joints in a volcanic tuff on Yim Tin Tsai (Plover Cove). These near-vertical joints act as planes of weakness that are opened up by weathering and erosion. Smaller joints and other cracks within the rock are responsible for the crisscross patterns on the rock surface.

IN09 Joints in Rock

Joints are smooth, planar fractures in rocks that extend from a few metres to several kilometres. Several types occur. Columnar joints result from cooling or shrinkage (IN40, p. 136). Sheeting joints develop parallel to the ground surface as erosion removes the overlying materials, thereby reducing the pressure and allowing expansion and cracking of the rocks. Usually these joints exhibit a gentle curvature.

Tectonic joints develop in response to earth movements, and are the most common, mainly occuring as a series of multiple, parallel, flat, intersecting cracks. They commonly control the shape of small rock outcrops. Excellent examples can be seen at Waterfall Bay on southern Hong Kong Island (top right and right), where three sets of joints intersect (figure below left). Note in the photographs how the joints (and a fault hidden behind the waterfall) control both the rock form and the overall rectangular shape of the bay.

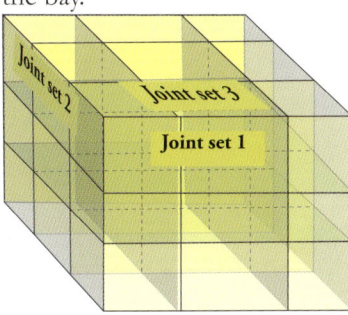

Three sets of parallel joints can be seen in the block (above left). Two are vertical and one is horizontal. All intersect at right angles. Note in the photograph and sketch (right) how similar joint planes appear in reality.

blocky, angular shapes in outcrops (IN09, p. 37). Joints mostly develop while rocks are buried under several kilometres of overlying materials. Their formation reflects the brittle nature of rocks and the stresses that are imposed on them by earth movements.

Similar to faults, joints are planes of weakness that influence the weathering and breakdown of rocks and, in turn, the development of landscape. They often intersect nearly at right angles, with streams, valleys, and coastlines commonly following these rectilinear patterns. Intersecting joints (IN09, p. 37) also add texture to valley sides and hilltops through the formation of blocky rock outcrops and the development of boulder fields (IN10, p. 39).

Boulder fields form where several joint sets (at least three or more) intersect (IN09, p. 37; IN10, p. 39). Commonly the rocks are also characterised by relatively large grain sizes and minerals, such as feldspars, that can be easily broken down chemically. The example above shows a boulder field formed from volcanic rocks on Luk Chau Shan in the northern part of the Ma On Shan Country Park. Boulder fields are usually best developed on granitic rocks. Excellent examples of the latter can be found on the western slopes of the Castle Peak Range in the western New Territories (below). In this area, abundant, well-rounded boulders are spread across many of the hillsides.

IN10 Boulder Fields and Joints

Boulder fields are loose accumulations of large, usually rounded, detached boulders that lie strewn across a land surface. They begin to develop when water slowly percolates down joints, weathering and weakening the original rock material on the edges of joint blocks and leaving a central rounded corestone of relatively undecayed material. The process that leads to the formation of corestones is called spheroidal weathering (adjacent photograph). When the rock is exposed at the surface, erosion begins to remove the surrounding loose rock debris, leaving behind fresh corestones over the land surface.

Stage 1. Fresh jointed granite or coarse-grained tuff. Rain seeps down intersecting joints and begins chemical weathering along these planes of weakness. Weathering is greatest at corners where the water can promote chemical breakdown from three sides. Weathering is moderately strong at edges, with only two sides to attack, and slowest in the middle of a block face. The result is that weathering gradually makes the fresh joint block rounder, producing the typical form of a corestone boulder.

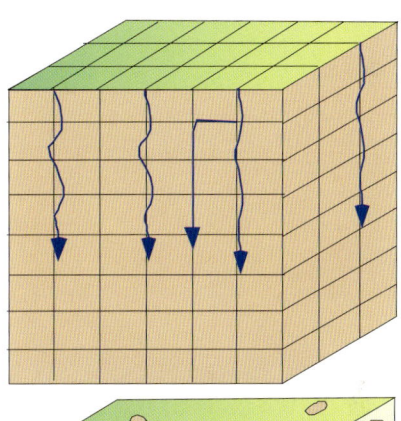

Stage 2. Chemical weathering causes rock to decay mainly by the breakdown of minerals such as feldspars to new minerals, typically clays. In Hong Kong, this is usually to a white clay called kaolin. As the process continues, rounded boulders develop that are more common and larger in the lower part of the profile. Isolated fresh boulders accumulate on the surface as the ground is lowered by erosion.

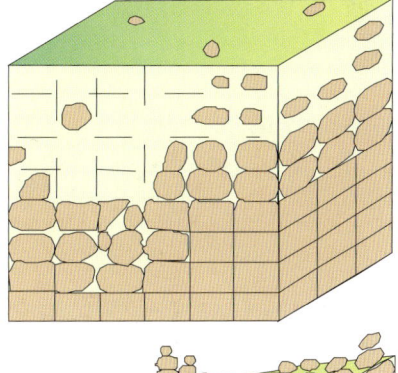

Stage 3. Further erosion removes most of the loose, weathered material. Exhumed corestones of fresh unweathered rock gradually accumulate on the surface, which by now has been considerably lowered. In some instances, boulders will be left resting on top of each other to form a feature called a tor.

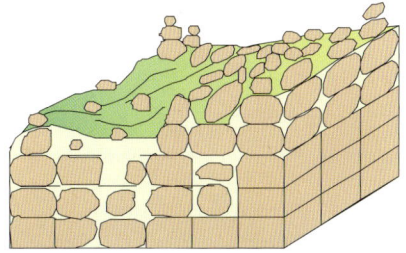

Coastal Landscapes

Tai Long Wan Cliffs,
Eastern Sai Kung

COASTAL LANDSCAPES:
CLIFFS, BEACHES, AND MUD FLATS

Coastlines reflect the interplay between geology (rock type, faults, and joints) and marine processes (erosion and deposition). These factors vary from place to place, resulting in the variable development of cliffs, beaches, and tidal mud flats along different sections of the coastline. Hong Kong's highly indented coast is about 350 km long and shows a marked contrast between west and east. The western shores tend to have a more subdued appearance (photograph below), with gentler slopes and either rocky or muddy coasts. In contrast, eastern areas tend to be much more rugged, with abrupt, near-vertical rocky cliffs (p. 40) and sandy bays.

The Western Region

Shorelines in this region are generally low-lying and depositional. Deep Bay, in the northwest, is very shallow, with extensive mud flats and fringing swamps, especially near the Sham Chun River mouth. An important mangrove community is established on the mud flats, with a rich and varied fauna, particularly of shellfish. Mangrove stems slow the tidal currents in the area, which promotes settling of the fine-grained suspended sediments and the seaward extension of the mud flats.

In contrast, the coastline adjacent to the granite uplands of the Castle Peak Range

Extensive tidal flats form the northwest coastline of Hong Kong. These can be seen just beyond the fringe of Mai Po mangroves next to the road. The area also includes fish ponds and gei wei wetlands where shrimps are harvested.

is comprised of a narrow, low-lying, coastal zone. Several sandy beaches, of varying size, alternate with accumulations of large rounded boulders derived by slope failure from the adjacent hillsides.

Lantau, the largest of Hong Kong's 262 islands, has a varied coastline (map below). In the north, volcanic and sedimentary rocks have produced rocky shorelines, with beaches and mudflats in bays protected from waves. During the last decade, much of the north has been altered by reclamation for the North

Small beaches (and a few larger ones) alternate with areas of large well-rounded boulders along parts of the southern Deep Bay coastline. The boulders originated on the neighbouring granitic slopes and have been moved to the shoreline by landslides.

Hong Kong coastlines vary with exposure to the sea, local rock types, and the degree of human intervention. The map shows the dominance of sea cliffs and steep rocky slopes in the southeast. Northwesterly shores tend to consist of steep rocky slopes plunging into the sea (without cliffs) or areas of mud- or sand-flats. Several types of wave-cut platforms occur in the northeast. Beaches are best developed on southeasterly facing shores. Artificial coastlines occur around Victoria Harbour, northern Lantau, inner Tolo Harbour, and the western New Territories. Place names indicate the locations of coastal features shown in photographs in this section.

Lantau Expressway. In contrast, the south is less developed. These coasts are exposed to stronger waves from the South China Sea and are characterised by rocky shores and extensive beaches, including Cheung Sha Beach, the longest in Hong Kong. In the southwest, granitic rocks (photograph below) have produced beaches and rocky, boulder-strewn coasts.

The Central Region

Hong Kong Island displays similar variety to that of Lantau. Most of the north shore has been reclaimed for urban development over a period of more than a century. There are several prominent peninsulas in the south, notably Stanley and Cape D'Aguilar, that have a gently sloping rounded form on their higher slopes. These inclines tend to become steeper lower down and, locally, may form sea cliffs. Commonly, large piles of boulders lie at the base, having fallen from above. Similar rounded hills and coastal

Lantau Island has been experiencing a radical change in its coastal landscapes since the development of the Chek Lap Kok Airport. In recent years, the northern shores have been significantly altered by infrastructure developments such as the expressway shown in the photograph above. Pressures on the area are increasing with several schemes for additional reclamation having been proposed.

The southwestern tip of Lantau, Fan Lau, consists of a remote, rocky, granite promontory, isolated beaches, and bouldery shorelines. Several islands belonging to mainland China can be seen along the horizon. These have suffered considerably from aggregate quarrying in recent years.

Coastal Landscapes: The Central Region

Victoria Harbour is surrounded by artificial coastlines and high-rise buildings and offers one of the most exciting human landscapes in the world.

cliffs are also present on neighbouring Po Toi, Beaufort Island, and Lamma.

The southern headlands on Hong Kong Island also enclose bays with extensive beaches, such as Sham Shui Wan (Deep Water Bay), Heung To Wan (Shek O), Tin Shui Wan (Repulse Bay), and Tai Long Wan (Big Wave Bay). The sand here owes its existence to the southerly facing coast, which has exposed these shores to strong ocean waves. However, in the past, major typhoons removed large volumes of sand from parts of these coastlines. The beaches were replenished artificially in the early 1990s.

Coastlines on southern Hong Kong Island consist of eroded headlands, and bays where deposition occurs. Waves move the eroded material to the head of adjacent bays, such as those at Shek O (below), where it contributes to the beaches. Most of the sand, however, is supplied by rivers.

44

The Eastern Region

Shorelines in the east and southeast are generally erosional. The coast is mostly made up of fine-grained volcanic rocks that, in some places, have well-developed vertical joints (IN09, p. 37; IN40, p. 136). The action of the sea on these rocks has created a distinctive coastline with high, precipitous cliffs, especially in the extreme southeast. For example, Bluff Island (photograph below) is rimmed by the tallest cliffs in Hong Kong, which reach over 140 m high.

In many places, the cliff faces are penetrated by sea caves. Sea arches develop where extreme erosion has occurred. Collapsed arches have, in a number of places, produced small isolated islands called stacks (figure, p. 138). Sea caves

The southeastern coastline is characterised by high and rugged cliffs, such as those shown above, on Wam Tam Shan (Basalt Island), and below, on Sha Tong Hau Shan (Bluff Island). The cliffs are particularly distinctive because of the numerous vertical columns that occur in this area. These originally formed by cooling of hot volcanic ash, which causes hexagonal vertical joints to develop (IN40, p. 136). Today, they are being attacked by powerful ocean waves, with the columnar shapes guiding the form of the cliffs.

are common along the coast and on many islets between Tai Long Wan and Basalt Island (eastern Sai Kung Country park). Unusually, Tai Chau (adjacent photograph) is penetrated by an underwater cave that runs the entire width of the island.

Rugged cliffs also occur in the extreme northeast of Hong Kong, such as on parts of Kat O. However, wave erosion on several other islands has cut gently sloping rock platforms into the sedimentary and volcanic rocks. Around Double Haven, most of these platforms occur slightly above the current high tide level (e.g. on Crooked Island, Crescent Island, and Double Island). Elsewhere, such as around Hok Tsui and Ping Chau, these platforms lie below the high tide level.

Tai Long Wan (above) contains three islands of volcanic rock. Lan Tau Pai is the very small islet to the left. Tai Chau is the large central island, and Tsim Chau lies to the right. An underwater cave runs below Tai Chau.

Crooked Island, or Kat O, has a finger-like shape. The photograph below shows Kai Kung Tau at the end of one such finger. Note the eroded cliffs on this exposed coastline. Wong Wan Chau, or Double Island (right), lies in more protected water. There, wave erosion and weathering processes have combined to generate flat platforms just above the high tide level.

PART THREE
HONG KONG REGIONS

The Southern Face of Lion Rock

The Northwestern New Territories

Landscapes characterised by rivers, lowland plains, and coastal wetlands impart a distinctive flavour to this region, providing scenery that does not occur in other parts of Hong Kong. Isolated mountain ridges also form a unique element of the landscape. Several of the mountains and valleys, and parts of the coastline, follow a NE–SW orientation, best shown by the Lam Tsuen Valley, Tai To Yan, and the coast at Lau Fau Shan. Human settlement is widespread, with villages scattered across the lowlands. New Towns are located at Fanling, Sheung Shui, Yuen Long, and Tin Shui Wai. A few older settlements occur at Shek Kong and Kam Tin.

The region includes the Lam Tsuen Country Park, which contains the high, steep-sided, volcanic ridges of Kai Kung Leng (572 m) and Tai To Yan (566 m).

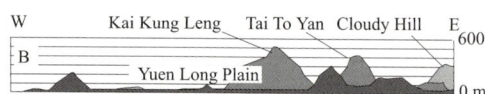

The northwestern New Territories extends from Shek Kong to the mainland border at the Sham Chun River. In a west-east direction, the area stretches from Deep Bay to Cloudy Hill. The two sets of profiles below show topographical cross-sections across the area. Dark shading represents foreground hills. Lighter shading represents land that lies progressively further away. Profiles in A run north to south, and are viewed from the west. Sections in B extend from east to west, and are viewed from the south.

Both Kai Kung Leng (looking west in the view above) and Tai To Yan are excellent ridge walks of 6–7 km. After steep initial ascents, they offer gentle hiking and open views of the surrounding plains from their grassy summits.

These grassy mountains separate the Fanling Plain and the Kam Tin River Valley. Cloudy Hill (440 m), in the east, is split from Tai To Yan by the NE-SW-trending Lam Tsuen Valley.

The panorama shows a southwest view from Kai Kung Leng. Yuen Long and Tin Shui Wai are situated across the Kam Tin River and are surrounded by lowlands and wetlands. The Castle Peak Range can be seen in the far distance.

The inappropriately named Deep Bay lies to the west and is less than 5 m deep. The waters are brackish due to the influence of the Pearl and Sham Chun Rivers. The bay is home to an oyster cultivation industry that uses floating rafts (adjacent photographs). After the oysters have been extracted, their shells are discarded on the coast near Lau Fau Shan. Thick, ridge-like middens have accumulated, giving rise to a unique coastal landscape.

Oysters, attached to ropes, are grown below rafts in Deep Bay (top right). After collection, they are spread on racks on land (left). The shells are discarded along the coast, and, over the years, a series of parallel ridges have developed (lower right, with the Deep Bay bridge under construction in the background).

Extensive wetlands occur at Mai Po and along the northwest coast, stretching over a 3-km-wide zone. Parts of the mud flats, mangroves, gei wei (artificial shrimp lagoons), and fish ponds, have been protected within the Mai Po Nature Reserve (map, p. 48), which is internationally recognised under the Ramsar Convention (IN13, p. 55). The area is a refuge for over 320 species of local and migratory birds.

Mai Po (above and left) includes a wide variety of habitats, ranging from tidal flats to freshwater ponds. Consequently, the area is an important sanctuary for birds and other animals. This ecosystem is recognised as being of international importance and is becoming of ever greater significance as coastal wetlands in the rest of southern China are threatened by development.

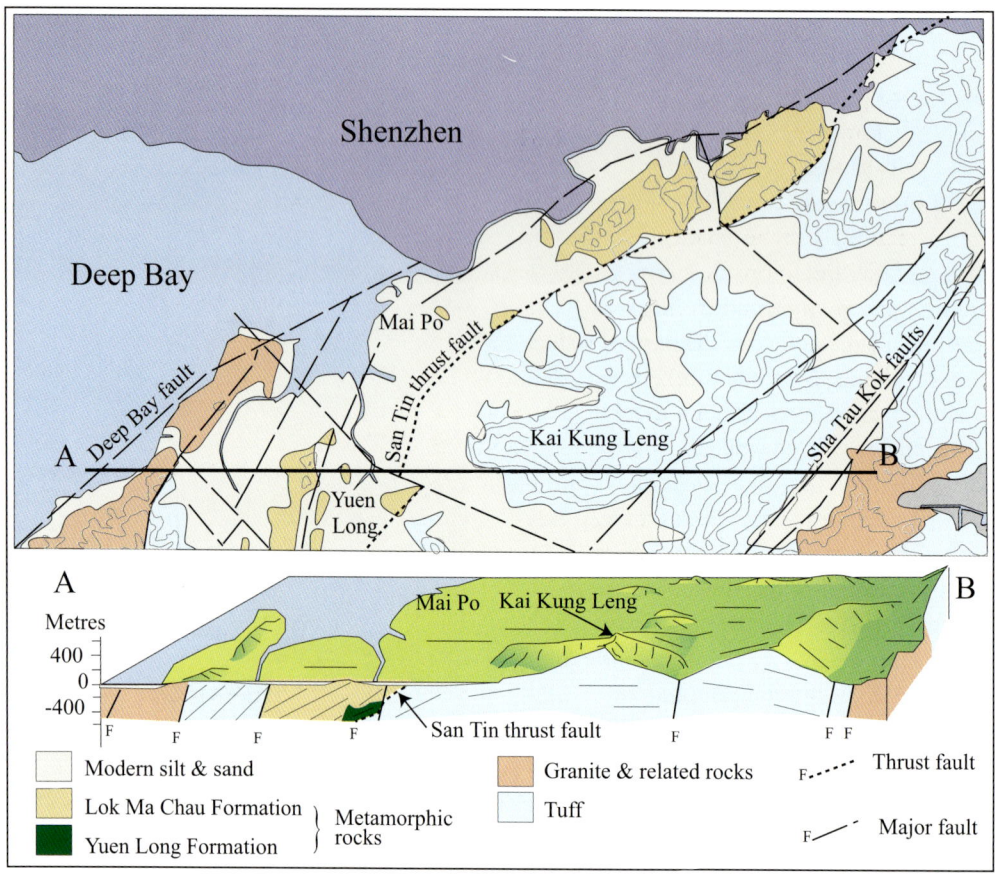

What Lies Beneath

Differences in rock type help to explain contrasting elevations. Hard rocks resist erosion and stand high above softer rocks. This is referred to as differential erosion (IN07, p. 31), and is clearly seen in this area, where resistant tuffs underlie the higher mountains, such as Kai Kung Leng and Tai To Yan. Granitic rocks form the lower, rounded hills in the southwest and southeast.

A band of metamorphic rocks (IN11, p. 51), confined between two faults, form a series of low hills. These rocks, which occur nowhere else in Hong Kong, extend southwards towards Tuen Mun and have an important influence on the subsurface topography, as described later. They were developed when heat and pressure changed the minerals and textures of pre-existing siltstones and sandstones to new rocks such as phyllite and schist (IN11). Marble (altered limestone) lies below the latter rocks. Though buried, marble has proved to be a problem during the development of Yuen Long and other towns such as Ma On Shan and Tung Chung. This is because

Both granite and granodiorite (IN04, p. 22) occur in the southeast and southwest of the region, but most of the area is underlain by thick sequences of volcanic rocks, mainly tuffs (IN02, p. 15). These extend below the silts and sands that have been deposited by modern rivers and in coastal settings. A NE-SW-trending band of metamorphic siltstones and marble (IN11, p. 51) occurs between two sets of faults. The cross-section above shows the rocks below the ground surface. The layering of the volcanic rocks is represented by long dashes in the cross section. They have been folded by earth movements and now dip steeply to the west in some areas and are nearly horizontal elsewhere. Most of the faults are nearly vertical, except for the San Tin thrust fault (IN08, p. 35), which is inclined gently to the west.

IN11 Metamorphic Rocks

Both pressure and temperature increase with depth below the ground surface. When subjected to increases in either or both of these factors, the type of minerals in rocks, their crystal sizes, and their textures can be changed. This takes place without melting to produce entirely new rocks that are described as metamorphic. The new minerals are usually plate-like and lie parallel to one another. This creates a layered texture referred to as foliation, which occurs in rocks such as phyllite and schist. In some instances (gneiss), alternating bands of light- and dark-coloured minerals develop that may be a few to many centimetres thick. Metamorphic rocks can be formed over extensive regions by burial during a mountain building episode, or they might develop locally close to hot molten magma.

Metamorphism is analogous to pressure cooking. Pressure and temperatures change the components of a recipe. However, by increasing the number of ingredients, the possible outcomes are expanded. Rocks with a simple chemistry tend to produce a limited range of metamorphic rocks. The sedimentary rock limestone (mostly just calcite), for example, forms marble. In contrast, mudstones have a complex chemistry because they contain a number of different minerals. As temperature and pressure increase, they form a great variety of new metamorphic rocks. However, although complex in detail, the new rocks can be categorised into just a few major types that develop in a particular sequence. The first to form from a mudstone would be shale, then phyllite, followed by schist, and finally gneiss (diagram below).

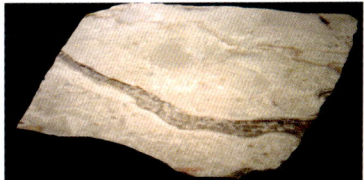
Marble—white with a sugary texture

Shale—grey, tending to split on flat plains

Phyllite—shiny, splits along undulating surfaces

Schist—shiny with a tendency to split along irregular surfaces

Gneiss—minerals are separated into alternating dark- and light-coloured layers

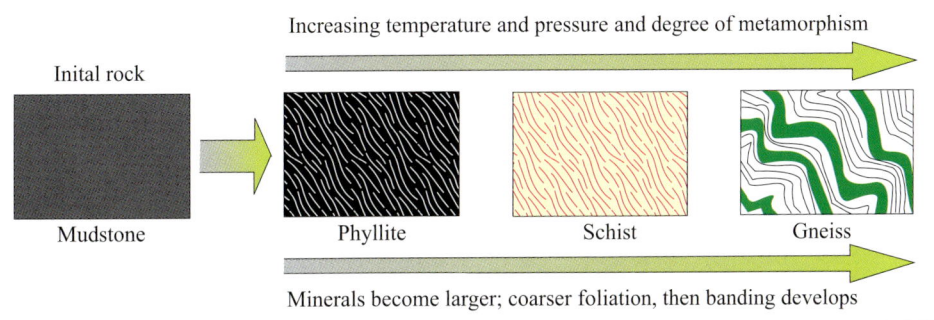

marble dissolves relatively easily in the slightly acidic waters that percolate through the rocks, especially where they are fractured by joints and faults. Consequently, cavities (partly filled with debris) have formed. These have proved to be a problem when constructing foundations below tall buildings, which need to be seated on solid rocks. Most importantly, the bedrock surface is highly irregular, with depressions developed where there are joints or faults, and pinnacles or ridges remaining over the intervening unfractured rock. This is termed a buried karst (karst, in general, refers to a landscape formed by rocks that dissolve).

Dissolution by acid groundwater has developed a karst surface on the buried marble below Yuen Long. The marble also contains a number of cavities, partly filled with sediment, which create problems for foundations in the area.

Two types of faults are common. Most are strike-slip (IN08, p. 35), with NE-SW trends that control valley alignments (e.g. the Lam Tsuen fault; map, p. 50). A few strike-slip faults follow a NW-SE trend and are responsible for valleys in this direction. The San Tin thrust fault has little topographical expression. It also trends NE-SW, lies at a low angle and has carried older metamorphic rocks over the top of younger tuffs.

In Hong Kong, folding (IN12) plays a less important role in controlling landscape than do rock types or faults. Nevertheless, two sets of folds occur in the northwest. The younger folds are broad synforms and antiforms (B in IN12) that have affected all of the rocks. The older folds only affect the ancient metamorphic rocks and consist of asymmetrical antiforms that have been broken by thrust faults (D in IN12).

IN12 Folds

Earth movements are common and rocks can respond in several ways, depending on the rate of movement, and whether the rocks are brittle or able to bend (ductile). Compression, or squeezing, of the rocks can result in reverse and thrust faults (IN08, p. 35), but may also cause rocks to be folded. There are many fold styles, but the simplest classification recognises upfolds (antiforms) and downfolds (synforms), which may be tight (A) or more open (B). If the rock layers in the centre of these folds become younger upwards they are called anticlines and synclines. Geologists recognise an imaginary plane (the fold axis; A) that bisects a fold, with the sides of the fold being referred to as "limbs" (A).

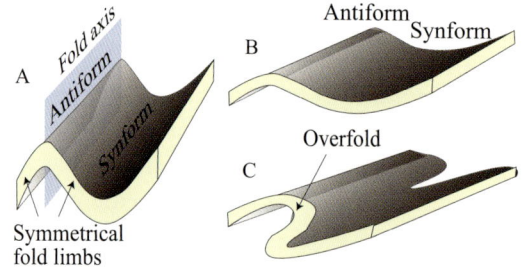

In some cases, the folding may cause a limb to turn upside down to produce an overfold (C). Sometimes, faults may cut through a fold, commonly along the fold axis, and offset the limbs (D).

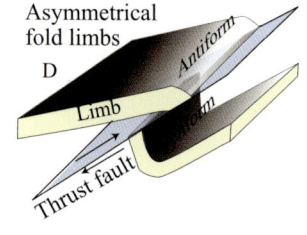

Resistant volcanic rocks form the higher ground in the northwestern New Territories. The summit ridge of Kai Kung Leng is dominated by coarse-grained tuffs, which are exposed along several sections of the main ridge line.

Tai To Yan is also underlain by volcanic rocks. The western end of the ridge is sharp and steep-sided. However, further to the east, the ridge is broad and undulating and surmounted by a number of small rounded peaks.

Rivers and Floodplains

About 5% of Hong Kong is occupied by rivers and their floodplains, mostly in the northwest. The Yuen Long and Fanling Plains, for example, are formed by sediments laid down by rivers. This deposition occurred mostly during the late Pleistocene and Holocene periods (the last few tens of thousands of years). Parts of the Yuen Long Plain are made up of buried marine sediments that were laid down at a time when a large bay occupied the area (IN13, p. 55).

Over many millennia, streams may erode mountainous areas to a flat surface. However, uplift, caused by earth movements, commonly interrupts this lowering of the surface and rejuvinates streams so they can cause greater incision. Initially, with greater elevation, streams are highly erosive and carve deep V-shaped valleys. These streams are referred to as youthful. With time, maturity is achieved and streams erode sideways to form small floodplains. In old age, they develop large loops (meanders) and create a much wider floodplain. Streams in Hong Kong exhibit all of these stages. For example, V-shaped valleys are common on the mountains, and the Lam Tsuen Valley is occupied by a mature stream.

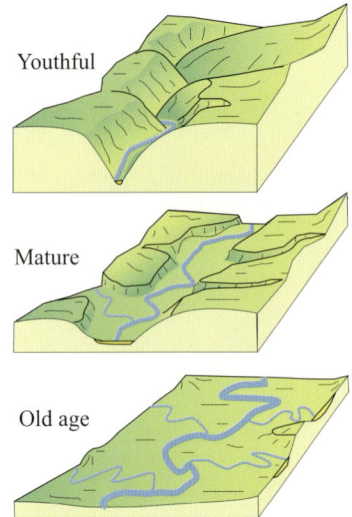

Erosion of mountains by streams follows a distinctive sequence. In the youthful stage, erosion is dominant. Streams cut downwards, carving out V-shaped valleys with interlocking spurs. In maturity, lateral erosion widens valleys and sediment is deposited on small floodplains. In old age, erosion is negligible and sedimentation dominates, with rivers winding across broad flat surfaces.

A youthful stream draining Tai To Yan (lower left) has carved a V-shaped valley. In contrast, the Tai Hang Valley (above) is occupied by a mature river, with a broad floodplain between Tai To Yan, to the right, and Cloudy Hill, to the left. The Kam Tin Floodplain (lower right), with Tai Mo Shan in the background, has been formed by rivers in the old age stage of development. The low relief surface was once crossed by several meandering rivers that have now been straightened artificially.

IN13 Mai Po and the Wetlands

About 18,000 years ago, when glaciers reached their worldwide maximum extent, the sea level was 120 m lower than today. Then, as a result of natural global warming, ice on land melted and water returned to the sea, which rose to modern levels by about 6,000 years ago. It may even have stood 2 m higher than today around 5,500 years ago. They have changed little since. At that time a bay covered much of the lowlands around Yuen Long and Kam Tin (map).

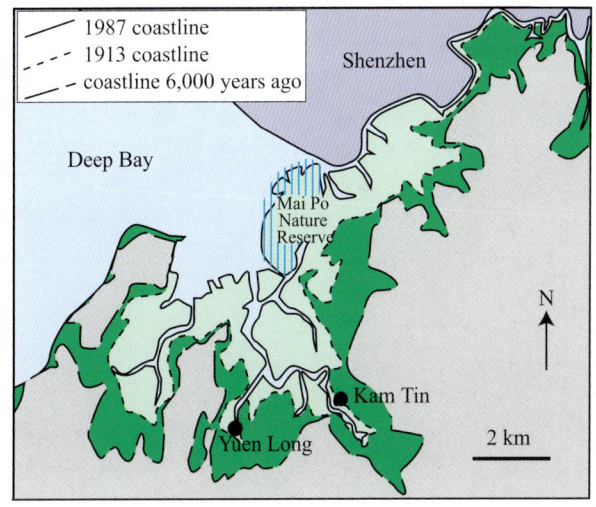

Although sea level has remained constant, the process of progradation caused the coast to advance to its 1913 position (map). Progradation occurs when river water slows down upon meeting the sea and the suspended particles settle to the sea floor, accumulate, and form new land. During the twentieth century, the coastline advanced naturally further seawards, but human activities played an increasingly important role. For example, extensive shrimp farming began in the mid-1940s after the construction of ponds (gei wei) on land reclaimed from coastal mangroves.

The Mai Po Marshes are now the sixth largest mangrove stand along the coast of China and include one of the most extensive reed beds in Guangdong Province. The marshes and wetlands are home to 12 endangered bird species and are regularly visited by over 20,000 wintering water foul. In January 1996, 68,000 birds were recorded in the Mai Po-Deep Bay area. The importance of the region as a wetland was recognised internationally when it was designated as a Ramsar site in 1995.

The Mai Po marshes and wetlands, just beyond Fairview Park, are internationally important bird sanctuaries.

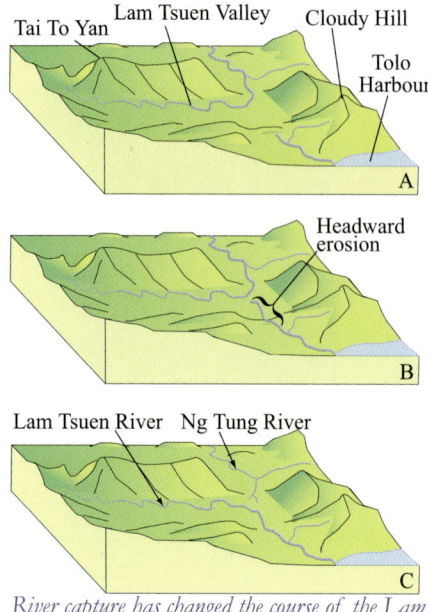

River capture has changed the course of the Lam Tsuen River, diverting the system towards Tolo Harbour.

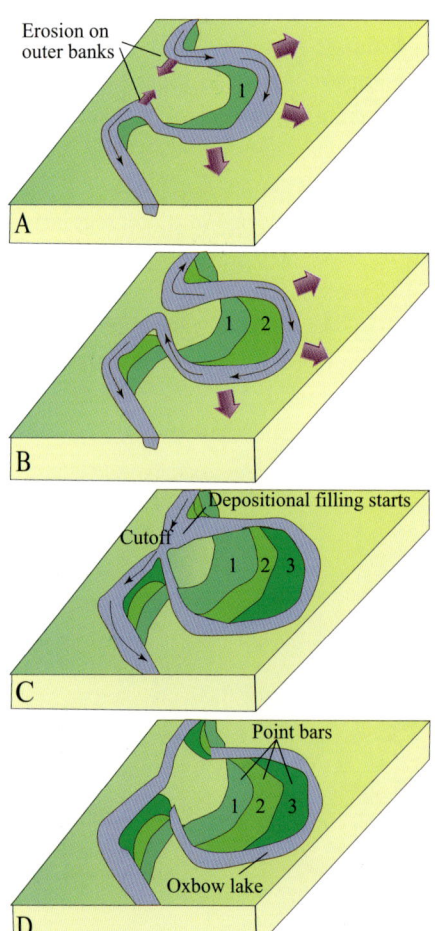

Streams can be rejuvenated by uplift, as noted above, or if the water they contain increases following changes in rainfall. A third process involves the diversion of water from another stream. This is called river capture, and appears to have occurred, for example, in the Lam Tsuen Valley. The stream there once flowed north to join what is now the Ng Tung River (A on left). However, another stream that flowed into Tolo Harbour may have gradually eroded its channel in a headwards direction (B) to intercept the former course of the Lam Tsuen River, diverting it southwards (C). A further mechanism involves faulting, which may have dammed the old river course and opened the valley towards Tolo Channel.

Rejuvenation can also be caused by changes in sea level. The sea is the lowest point (base level) to which streams can erode—streams cannot flow uphill to the sea. When sea level falls, as has happened several times over the last few hundred thousand years, streams begin to cut down to a lower base level, incising into the existing deposits. This process results in the formation of river terraces (p. 58).

Floodplains are flat landforms created by meandering rivers through processes that occur both within and outside the channel. Streams shift position over time as a result of erosion of the outer bank, where water flow is fastest, and deposition on the inside of the bend, where the stream velocity is slowest. Erosion on the outer bend carves out the valley, whereas deposition on the inside bend forms a point bar of sand and gravel (IN14, p. 57). During floods, water overflows from the channel and slows down,

The figure shows the process of meander formation and abandonment. Erosion proceeds rapidly on the outer banks, where the water flows fastest (A). The accumulation of sediment forms point bars (1–3) on the inside bend. This process continues with the two bends gradually getting closer until they meet at a cutoff point (B–C). Sediment plugs and cuts off the former meander, and the channel is abandoned to leave an oxbow lake. River channels repeatedly increase their meandering and then straighten themselves through this process.

IN14 Floodplain Sediments

River deposits are called alluvium. During floods, water overflows from the channel, slows down, and loses its ability to carry sediment. Sand, being heavier, settles first, forming a ridge called a levee (B) on both sides of the channel. Smaller particles of silt and clay, being lighter, are carried further. Eventually, the water slows sufficiently for the silt to sink, and, then as the water stagnates, clay is dropped. Thus a graded couplet of silt/fine sand overlain by a layer of clay/silt is formed (G). Each couplet represents a single flood. In Hong Kong, roots, termites, and other organisms destroy the layering, a process called bioturbation.

Coarser sediment is deposited within channels as point bars (D–F). The velocity of the water flowing around a bend varies, being fastest on the outside and slowest on the inside. Consequently, the river erodes its outer bank, causing the channel to migrate sideways. On the inside, the stream flows slowly and sediment is deposited as a point bar. This velocity gradient is reflected in grain size, which decreases towards the inner bend. As the bar enlarges and the channel migrates sideways, the stream leaves behind a layer of alluvium (H) with coarser basal deposits (mid-channel pebbles) that give way upwards to finer sands laid down on the inner part of the meander.

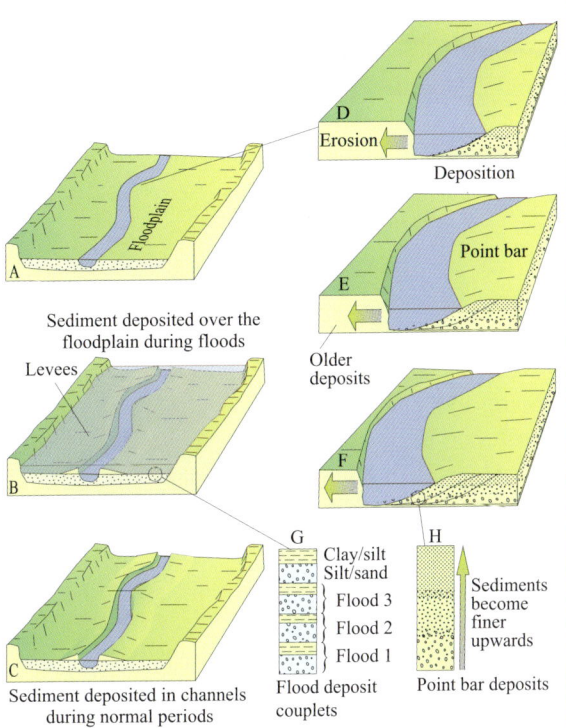

Both modern and ancient alluvium can be seen in this photograph. Alluvium is forming today at the point bar to the lower left. Finer-grained silt and sand is being deposited on the inside of the bend, and coarser pebbles have accumulated towards the stream centre. Note that, at the outer bank where stream flow is fastest, the stream is eroding into older alluvium. The ancient alluvium in the cliff is made up of pebbles and sands that were originally formed in a similar way to the modern point bar. Note also how the sediments become finer-grained upwards, with pebbles and boulders giving way to silt and sand. This is the typical pattern for alluvium formed by meandering rivers on floodplains.

depositing coarse sediments near the channel and forming raised banks called levees (IN14, p. 57). As the floodwaters spread, they slow further, and finer-grained silt and clay settles onto the floodplain.

An example of the development of a set of terraces is shown to the right. An initial valley filled with river sediments (alluvium) is shown (A). A fall in sea level or increase in river discharge (from heavier rainfall) causes the stream to cut downwards, leaving the old floodplain standing above the river (B). The confined stream then begins to meander and erode sideways to form a new floodplain within the older floodplain, separated from it by steep bluffs (C). The remnants of the older surfaces are called terraces. In this situation, they are referred to as paired because they lie at the same height and are fragments of the same former floodplain.

Although they are seldom noticed under modern developments, terraces occur on both the Yuen Long and Fanling Plains, and also in the Lam Tsuen Valley (p. 59) and Kam Tin Valley. These river terraces are ancient floodplains formed during the late Pleistocene period (between 80,000–20,000 years ago). Rivers cut into these older surfaces as the sea level dropped many tens of metres. Subsequently, during the last 10,000 years, new floodplains developed within the older sets. The heights of the Pleistocene terraces on the Yuen Long Plain, for example, range from

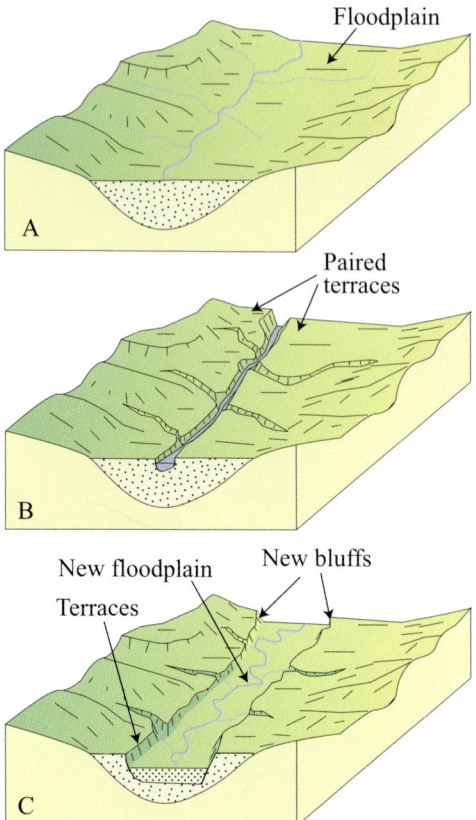

Paired terraces are created when a rejuvenated stream cuts down into its floodplain (B). Eventually, a new floodplain develops, inset within the older one (C).

just below sea level, at about −2.5 m to as high as +23 m.

The processes of stream incision, terrace formation, channel meandering, and levee development add form and shape to otherwise apparently featureless floodplains. In truth, these are dynamic, and not the inert and characterless landforms that they are commonly perceived to be.

The flat, low-lying ground around Fanling, Sheung Shui, and areas to the north are part of an extensive floodplain system that was formed largely by the Ng Tung River. The photograph above was taken from Lung Shan, a small but distinctive peak just north of Hong Lok Yuen.

The Northwestern New Territories: Rivers and Floodplains

The Lam Tsuen Valley contains many ancient terraces above the present-day river. These are old Pleistocene surfaces—a few tens of thousands of years old. Generally, they are difficult to see because of extensive trees and buildings. The photograph above shows one terrace, obscured by trees, to the right of the river, that stands above the modern floodplain. The left photograph shows a clearer example with steep bluffs. The modern river is extensively altered by human impacts except at a few short sections such as that shown in the photograph below.

Human Impacts

Deforestation and Erosion

The mountains, valleys, and floodplains of Hong Kong were once covered by extensive forests. These have changed through time as a result of natural climate variations. About 20,000 years ago, when glaciers covered much of Europe and North America, Hong Kong was dominated by semi-arid, open, temperate woodlands of coniferous or broad-leaved trees. The coastline at that time lay about 120 km to the south, exposing a broad flat plain that was covered by tall, dense grasses.

From about 18,000 years ago (figure below), the climate slowly warmed, glaciers melted, and the vegetation evolved. Tall tropical rainforests, previously confined to the islands of Southeast Asia, expanded and covered most of southern China by about 9,000 years ago. Subsequently, these forests retreated slightly and would still cover Hong Kong, if not for extensive clearances for agriculture and damage caused by accidental and deliberate fires.

The earliest human effects on these rich forests were probably exerted by coastal dwellers.

The natural vegetation of Hong Kong should be tropical rainforests. However, due to agricultural activities and fire, the mountains and valleys are instead dominated by grasslands and secondary forests. Grasslands tend to occur on highlands and mountain summits (top) and are maintained in this condition today by accidental fires. Woodlands dominate on the lower slopes and rise towards the peaks along stream valleys (bottom).

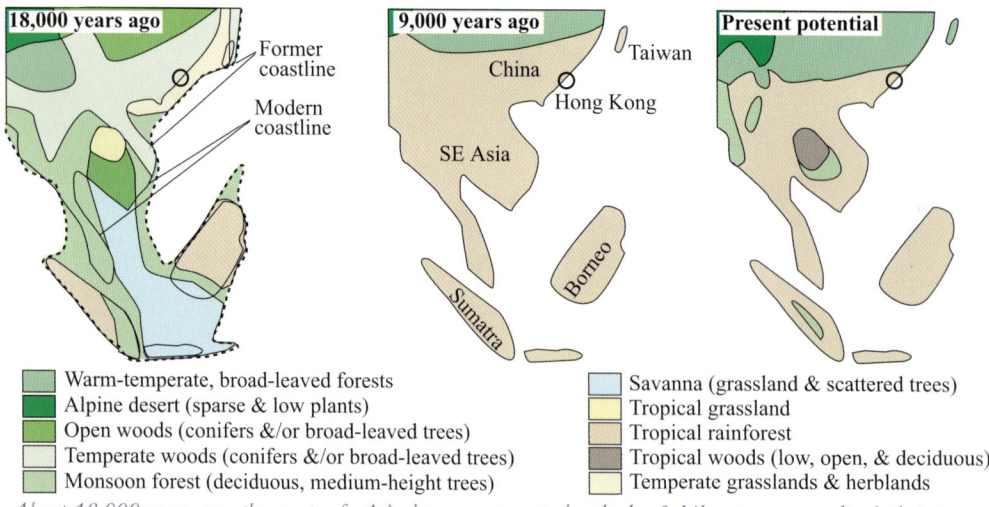

About 18,000 years ago, the coasts of Asia lay many tens to hundreds of kilometres seawards of their present positions. Southern China was dominated by extensive temperate woodlands, with grasslands occupying areas that are now flooded by the sea. By about 9,000 years ago, tropical rainforests had expanded across much of southern China. The present potential map (above right) shows how vegetation would be today, if not for human influences.

Archaeological evidence suggests that settlements were established along the shoreline of Hong Kong between 6,000 and 4,250 years ago. However, at that time, the population centres were small and their effects were limited to a few local settings. This situation was destined to change dramatically.

Evidence for human influences can be found in mud (in cores from Deep Bay), which preserves a record of environmental instability, both in the sea and on land. The relationship between environmental changes on the land and in the sea is very close. For example, a loss of tree cover will result in increased erosion on mountain slopes. Rivers transport the eroded materials to the sea, where they are deposited. Consequently, increased erosion on land results in increased deposition at sea. Similarly, deforestation changes the nature and supply of organic materials and nutrients to coastal areas, which can then cause shifts in the composition of marine flora and fauna. Analyses of the mud of Deep Bay suggest that major changes on land (related to agriculture and deforestation) had occurred by the late fifteenth century.

Documentary records of population movements in the south China region endorse these conclusions from the geological evidence. Guangdong became a part of the Qin Empire in 214 BC, but the Han Chinese did not settle there in large numbers for another thousand years. The early settlers belonged to the Five Great Clans that are still present in the New Territories. The Tang clan, for example, left Jiangxi and emigrated to Guangdong in the tenth century, and then travelled on to what is now called the the New Territories in 973 AD. They arrived in an area that was said to contain "luxuriant forests, verdant hills, and clear springs". Their early agriculture depended on rice, sweet potatoes, and sugar cane, which were produced in abundance.

The population of Hong Kong remained small until at least the twelfth century. By the thirteenth century, the initial immigrants (the Punti) occupied most of the rich lowland valleys and rented hillsides to later Hakka settlers. The development of farming on mountains and hillsides would have resulted

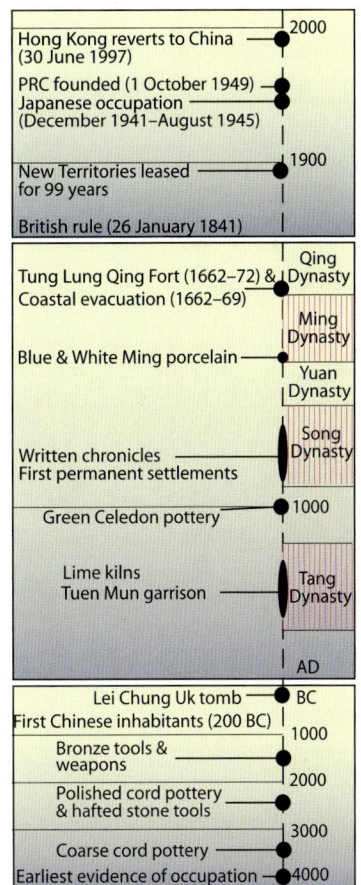

Original forests have been replaced with secondary growth (above) as a result of human occupation. Archaeological data and written chronicles suggest that settlement took place over several millennia (below). Agriculture, from the Ming Dynasty onwards, probably played a significant role in loss of the tree cover.

in significant deforestation and accelerated erosion. Immigration continued into the fourteenth century, when the Tanka and Hoklo arrived. They initially lived on boats in sheltered inlets, and later in coastal villages. Thus, the historical records provide firm evidence for increasing human impacts on landscapes, with agricultural expansion leading to the changes that are reflected in the mud of Deep Bay.

Boat people started to arrive in Hong Kong during the fourteenth century. Today many continue with their distinctive way of life.

Changing Coastlines, Stranded Villages

By the later part of the nineteenth century, approximately 100,000 people lived in the New Territories in about 800 villages, including wai (walled) and tsuen (not walled) settlements. Over half of the population was Punti (Cantonese-speaking), while the remainder were mainly Hakka.

The earliest reclamations of land from the sea in Hong Kong were of the extensive tidal flats that fringed the bays and estuaries in the New Territories. Their

The Mai Po area was reclaimed from a series of coastal swamps, marshes, and mangrove stands at the beginning of the twentieth century. The newly formed land was mostly used for aquaculture (shrimps and fish), and the results can still be seen today (below) in the form of numerous ponds that dominate the landscape.

primary purpose was to increase the area of agricultural land. Reclamations at Sha Tau Kok, Nam Chung, Luk Keng, Shuen Wan, and Yuen Long were accomplished by erecting bamboo stake fences, along which banks of marine mud were constructed. Landward of these barriers, the saline intertidal muds were desalinated using freshwater that was regularly flushed in through a system of sluices. Following a slow process of leaching over 7 years, the generally accepted period of treatment, the muds were finally considered to be suitable for cropping. During cultivation, continued ploughing and planting of the reclaimed mud resulted in the formation of a plough layer, a stiff clay near the surface. In order to maintain the fertility and productivity of the fields, this layer was periodically skimmed off and used to make the distinctive bluish-grey bricks that characterise many of the traditional village houses in the northwestern New Territories.

Between 1900 and 1950 there were many local reclamation projects and changes in landuse in the region to the north of Yuen Long. The first large reclamation was carried out in 1919 by a group of farmers who formed a company to reclaim 1,200 acres of salt-water marshes near Ping Shan. They also erected the village that is still occupied today.

Recognising the success of this major project, residents of other villages began negotiations to attempt larger reclamations around Mai Po in 1920. However, the proponents encountered difficulties when residents of Kam Tin, which lay upstream of the planned project, objected in the belief that the ponding works would severely disrupt the existing natural stream course, making their land more prone to flooding. Heated negotiations continued for 2 years until, in 1922, a compromise was agreed to, thus avoiding a potentially protracted feud. Following a revision of the plan, the Mai Po section of the reclamation was successfully completed in 1923.

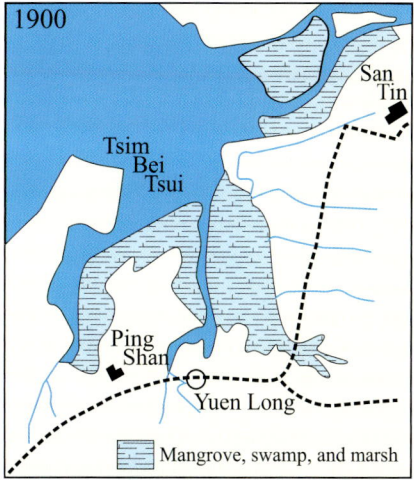
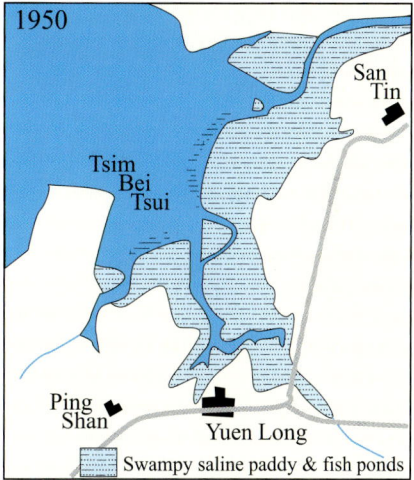
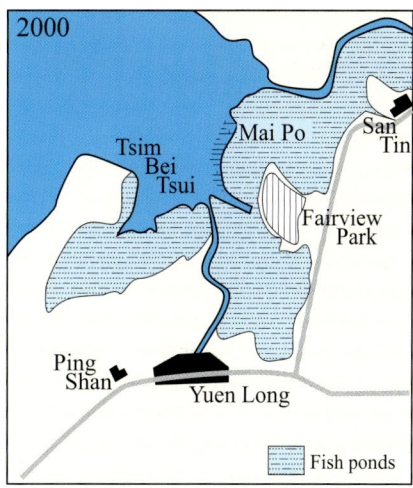

Reclamations during the twentieth century changed the shape of the coastline and the nature of the landscape, with the development of gei wei (ponds) in an area that was once occupied by a series of tidal swamps and mangrove stands.

A boom in land values during this period spawned a myriad of similar reclamation plans, but a recession in 1924 resulted in these plans being abandoned. Although other smaller reclamations, scattered across the Yuen Long marshes, were completed in subsequent years, none equalled the extent of the first ambitious project in 1919.

In addition to the artificial reclamations, natural reclamations are slowly taking place in the waters of Deep Bay. For example, the outer edge of the Mai Po Marshes Nature Reserve has extended seawards by 7.6 m a year since 1949. Consequently, the area of mangrove forest has trebled since the construction of the gei wei (shrimp) ponds and their enclosing mud embankments over 50 years ago. Average seaward rates of mangrove extension of 11.5 m a year were recorded between 1949 and 1969, and of 3.4 m a year between 1969 and 1987. Mud was derived both locally from the Shum Chun River and from the Pearl River that flows past the western entrance of Deep Bay.

The Demise of Hong Kong Rivers

Over the last two decades natural floodplains (p. 65) have been increasingly submerged under numerous container storage yards, new towns, and a rapidly expanding transport infrastructure. These developments have had the unintended effect of reducing the natural flood storage capacities of the area, causing more frequent and severe flooding problems.

As a result of these human impacts on the area, the Hong Kong government embarked on a major flood prevention programme that has involved both the widening and straightening of natural stream channels.

Deep Bay extends over an area of 112 km² and is fringed by extensive intertidal mud flats that support dense mangrove forests. These forests, co-dominated by Kandelia candel *and* Avicennia marina, *extend to 110 hectares on the northern shore and 90 hectares on the southern shore. The view in this photograph is southwest towards Tsim Bei Tsui (the background hill), with the tall buildings of Tin Shui Wai to the left.*

Mangrove leaves are specially adapted to survive the dry, low-tide periods in the harsh environment in which they live. Many have leaves with sunken stomata, epidermal hairs, and scales, as well as a thick waxy cuticle. These features all minimise water loss through evaporation. Furthermore, many mangroves also possess special water storage tissues.

Two views of the floodplains near Sheung Shui are shown in the photographs above. These subdued landscapes have been radically changed in recent years by extensive urbanisation and infrastructure development. Most of the former paddy fields have now disappeared, and the winding streams (marked by arrows in the upper photograph) have mostly been straightened and lined with concrete or stones to form sterile-looking nullahs.

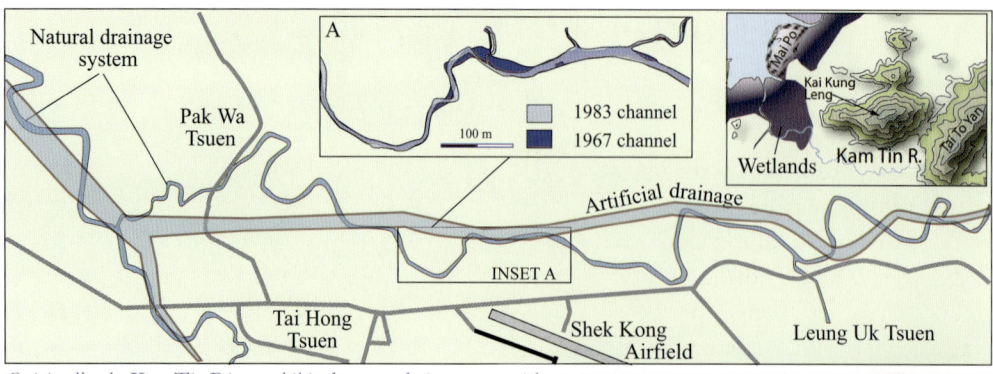

Originally, the Kam Tin River exhibited a meandering course with well-developed point bars (sand deposits on the inside of a river bend) (adjacent photograph). Extensive flood prevention measures have now radically altered the original course of the river by lining it with concrete or stones and by straightening and widening the channels to form artificial nullahs (lower right photograph).

The original meandering rivers have now disappeared and have been replaced by straight, concrete-lined channels. For example, about 20 km of sterile waterways have been constructed to protect Kam Tin and Yuen Long. The Ng Tung River, near the border with Shenzhen, has also been tamed with artificial nullahs. These concrete or stone-lined channels are designed to cope with the runoff from severe storms that might occur once in 200 years. Despite attempts to make them more attractive by adding boulders, grassing the banks, or adding earth linings, most remain unnatural and unsightly.

Nullahs reduce the risk of flooding by providing a larger channel cross-section and a straighter course. The former accomodates a greater water volume. Straightening also has an important effect. A stream that may have formerly meandered over 1 km for a fall of 1 m in height will then flow over a shorter distance for the same 1 m drop. In other words, the stream water will flow faster down a steeper gradient. The smooth man-made channels also minimise friction, which further increases the speed of flow.

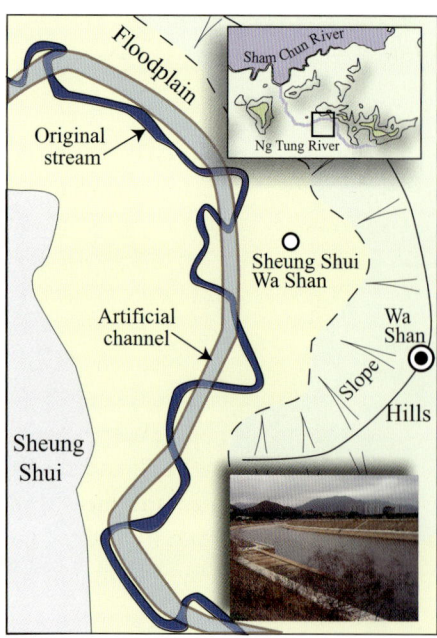

The altered Ng Tung River course provides another example of human impacts on Hong Kong's rivers.

THE NORTHEASTERN NEW TERRITORIES

Rocks exert a particularly strong influence on the shape of the landscape of the northeastern New Territories, which is distinguished from other regions by the extensive exposure of sedimentary rocks. The scenery developed on these materials is dominated by broadly E-W-trending ridges, some of which (e.g. Wang Leng) are asymmetrical in cross-section. These are characterised by gently inclined northern slopes and rugged south-facing cliffs. Others, such as the Bluff Head ridge with a NE-SW trend, are rounded and symmetrical. The highest mountain, at 603 m, is Wong Leng. This summit, other peaks, and the southerly slopes tend to be grassy, contrasting markedly with the woods that occur on northern gradients and along stream courses.

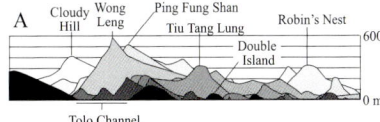

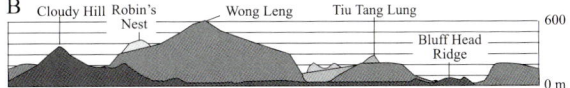

This region extends from the Tolo Channel in the south to Starling Inlet in the north. Asymmetrical ridges are common. This can be seen in the cross-sections shown below. Diagram A shows the region viewed from the east. Note the gentler northern (right) inclines and the steeper southern aspects of the larger mountains. From the south (B), the east-west extent of the ridges can be seen.

The region includes the Pat Sin Leng and Plover Cove Country Parks (IN28, p. 112), and the Closed Frontier Area (p. 77). It encompasses a variety of contrasting coastlines, ranging from mangrove-fringed bays to rocky promontories and idyllic island beaches. The shorelines of Tolo Channel and Starling Inlet (Sha Tau Kok Hoi) are comparatively straight because each follows a major NE-SW-trending fault line (geology map, p. 69). The northeastern coast is highly indented and faces a series of rugged islands around Double Haven. Humans impacts are significant, including, for example, the construction of the Plover Cove Dam, which blocks a former marine bay.

The symmetrical Bluff Head Ridge consists of a series of rounded hills that lie between Tolo Channel (right) and Plover Cove Reservoir (left). The ridge is underlain by sandstones, siltstones, and conglomerates. These have been tilted vertically by earth movements and today control the shape of the ridge.

There are no major centres of population. Over most of the more remote areas, hamlets and villages are largely abandoned. Thriving settlements occur at Wu Kau Tang, Luk Keng, and along the coast to the south of the Pat Sin Leng. The border town of Sha Tau Kok lies on the northern coast of Starling Inlet, but is not accessible without a permit.

The area includes some of the finest hiking trails in Hong Kong, notably the undulating and strenuous Pat Sin Leng (part of the Wilson Trail), Ping Fung Shan, and Tiu Tang Lung ridges. The distant parts of a circular walk around Plover Cove Reservoir offer some of the most remote and rugged walking in Hong Kong. More gentle hiking opportunities are found along the valleys; uplands such as Sha Lo Tung; and on the gentler northern mountain slopes.

Landscape Foundations

The roots of most landscapes lie in the geology that underlies a region. This is strikingly true for northeastern Hong Kong. The region is dominated by tuffs (IN02, p. 15) and sedimentary rocks (map, p. 69) that include siltstones, sandstones, and conglomerates (table, p. 27). These

The high mountain tops and ridges of the northeastern New Territories offer wide open vistas in all directions. The image below shows the Sha Lo Tung Valley from Wong Leng. The lower parts of the valley are famous for the abundance and diversity of its dragonflies. There are 72 species, representing almost 67% of the total number present in Hong Kong. The area is a home for four endemic species and provides a breeding site for six internationally rare dragonflies.

Many abandoned villages, such as this one at Kop Tong, occur throughout the northeastern New Territories. Most contain remnants of family life during the 1960–70s.

The Ten Highest Peaks (metres)	
1. Wong Leng	603
2. Shun Yeung Fung	591
3. Lai Pek Shan	550
3. Hsien Ku Fung	511
4. Robin's Nest	492
5. Hung Fa Chai	489
6. Kwai Tau Leng	486
7. Cloudy Hill	440
8. Tiu Tang Lung	416
9. Fan Kei Tok	369
10. Wang Leng	311

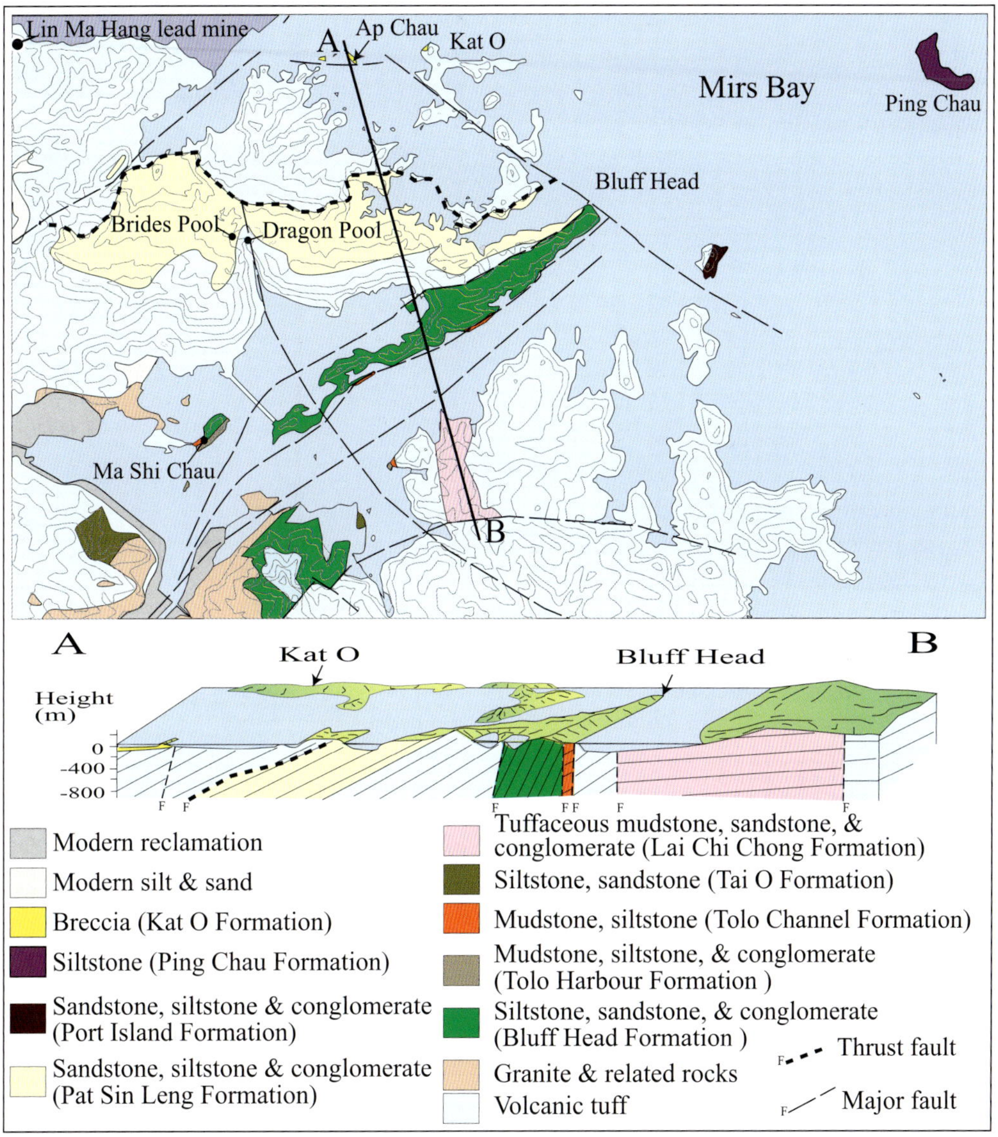

The Northeastern New Territories: Landscape Foundations

Modern reclamation	Tuffaceous mudstone, sandstone, & conglomerate (Lai Chi Chong Formation)
Modern silt & sand	Siltstone, sandstone (Tai O Formation)
Breccia (Kat O Formation)	Mudstone, siltstone (Tolo Channel Formation)
Siltstone (Ping Chau Formation)	Mudstone, siltstone, & conglomerate (Tolo Harbour Formation)
Sandstone, siltstone & conglomerate (Port Island Formation)	Siltstone, sandstone, & conglomerate (Bluff Head Formation)
Sandstone, siltstone & conglomerate (Pat Sin Leng Formation)	Granite & related rocks
	Volcanic tuff

The sedimentary Bluff Head and Pat Sin Leng Formations exert a strong influence on the landscape. The near-vertical beds of the Bluff Head Formation form a symmetrical ridge, bounded by faults to the north and south. In contrast, the gently inclined layers of the Pat Sin Leng Formation tend to produce asymmetrical ridges.

sediments formed at different periods and in contrasting environments. The oldest (400–360 million years) and most ancient in Hong Kong belong to a geological unit referred to as the Bluff Head Formation (IN06, p. 29). This comprises rocks that were originally laid down as loose silt, sand and pebbles in deltas and rivers near the sea. After burial under later materials, the sediments were compressed into rock,

At Bluff Head, light-coloured sandstones and dark mudstones have been tilted vertically. This orientation controls the shape of Bluff Head Ridge.

Many of the mountains of northeastern Hong Kong are asymmetrical, with gentle northern slopes and steep cliffs to the south. The left photograph shows the south face of Ping Fung Shan, near Wong Leng. Conglomerates and sandstones overlie tuffs on the cliff face. Further east lies another asymmetrical ridge between Tiu Tang Lung and Fan Kei Tok (centre). This ridge has a major thrust fault (IN08, p. 35) contouring along its lower slope. The thrust has carried tuffs over the top of sandstones and conglomerates (SS on photograph) (also see the thrust fault on the cross-section, p. 69). The third photograph shows Wang Leng, with steep cliffs on the southern slopes above Plover Cove Reservoir. Sandstones and conglomerates form the upper cliff line with tuffs below.

before being folded and tilted by earth movements. Today, the originally horizontal layers stand vertically, forming the narrow Bluff Head Ridge.

The second group of sedimentary rocks belong to the Pat Sin Leng Formation. This consists of younger (100–80 million years) sandstones, siltstones, and conglomerates that were formed in an arid landscape crossed by highly seasonal rivers (IN06, p. 29). These rocks and the tuffs that also occur in the formation have been less severely affected by earth movements than the older Bluff Head Formation. Their layers slope gently to the north and determine the shape of the many asymmetrical mountains in the region (IN03, p. 18; map, p. 69; photographs above), with steep cliffs cutting across the layering and gentle "dip slopes" that follow the sedimentary beds.

Layered rocks underlie several islands in Mirs Bay. Examples can be seen on Ap Chau (below.), and on Chek Chau (Port Island) (inset). The latter consists of layers of red sandstones originally deposited by rivers that crossed an arid landscape about 100 million years ago. In contrast, the layers on Ap Chau are made of large angular blocks that accumulated at the base of a scarp about 70 million years ago. On both islands, the shape of the land surface reflects the layering in the rocks.

Streams and Waterfalls

Numerous streams drain the mountains of the northeast. Those that follow the dip slopes (inclines parallel to the rock layers) tend to be longer, drain northwards, and have gentler gradients. In contrast, most of the streams that flow southwards cut across the rock layers and tend to have steeper gradients.

The floors of streams crossing the layered sedimentary rocks (mudstones, sandstones, and conglomerates) have a distinctive "blocky" appearance, since the rocks tend to be eroded along joints (IN09, p. 37) and bedding planes (IN05, p. 26). In several places, potholes occur in the stream channels. These depressions range from a few centimetres to a metre in diameter, although most are less than 10 cm deep. Potholes are formed by the grinding action of pebbles carried in turbulent water along the bed of the stream. These pebbles abrade the rock as they are swirled around during floods. Pebbles may be found in some of the depressions. Potholes commonly form in moderately resistant rocks, such as the sandstones and conglomerates.

There are several notable waterfalls in the region. The two largest are at Brides Pool and Dragon Pool (map, p. 69). Waterfalls can form as a result of a fall of sea level (IN19, p. 95). However, these two examples are the result of differential erosion (IN07, p. 31) of rocks. Brides Pool waterfall drops about 15 m over a hard and erosion-resistant conglomerate. The rocks below are mainly shales that are more easily weathered and eroded, thus they have been worn away over the millennia. As the softer layers are undercut, harder rock above forms a ledge that projects further

Bedding planes are the boundaries between layers of sediments and represent a time break—a period of non-deposition. These form planes of weakness after the loose sediments have turned into solid rock. The photograph above shows several bedding planes sloping gently to the right. Two joint planes are shown in the photograph. Joints develop as the rock is formed or by later earth movements. They are also planes of weakness that are picked out by erosion and control the shape of this stream bed, giving it a blocky appearance.

Potholes are depressions with a circular bowl-shaped form. They are carved out by pebbles caught in swirling eddy currents. Note also the rock wall at the back of the photograph. This is a vertical joint face. The gently sloping surface cut into by the pothole is a bedding plane. Several thin, white, quartz veins are also visible.

The Northeastern New Territories: Streams and Waterfalls

and further until, lacking support, it collapses. By this process, the waterfall has retreated up its valley. At the bottom of the waterfall, a large depression (a plunge pool) has been carved out. The pool, which is 25 m across and 2 m deep, was named, according to legend, after a bride who was being carried to her groom's home in a closed sedan chair. One of the carriers reputedly slipped as they crossed the stream, and the bride was swept to her death over the falls.

Brides Pool was carved by falling water crossing a hard layer of rock. It is an attractive cool place to visit but has a tragic local mythology.

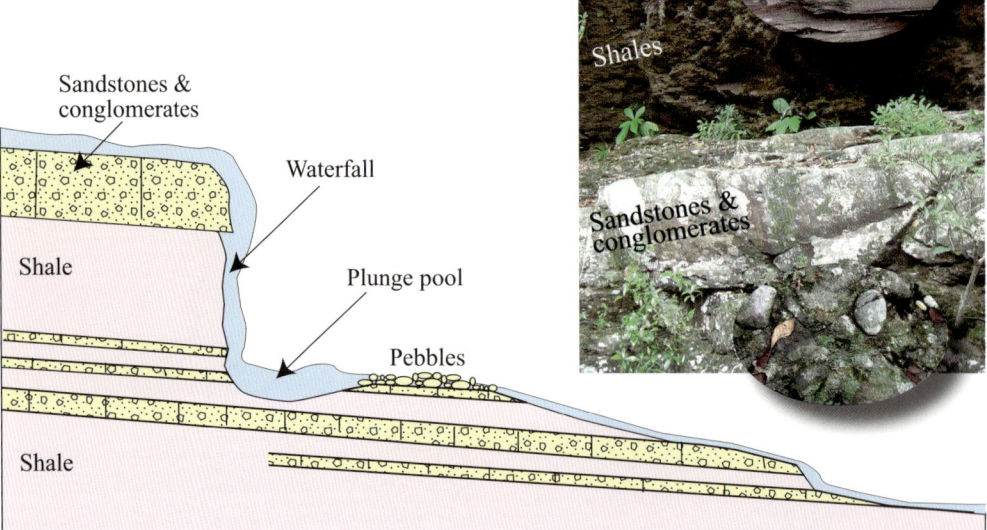

The Brides Pool waterfall was formed by undercutting of softer shales that lie below hard and resistant conglomerates and sandstones. Examples of shales, sandstones and conglomerates are shown in the photographs on the right.

The Northeastern New Territories: Streams and Waterfalls

This attractive waterfall lies between two pools. At the base is the Dragon Pool, which lies about half a kilometre to the east of Brides Pool. Mirror Pool (Chiu Keng Tam) lies just above the falls (about 20 m high).

Marine Inlets and Islands

The area is bordered to the north and south by two long marine inlets: Starling Inlet and the Tolo Channel (photographs, p. 75). Both were river valleys about 10,000 years ago. These valleys were flooded between 8,000 and 6,000 years ago, when sea level rose as glaciers melted at the end of the last Ice Age.

The northeastern part of the area has a highly indented coastline with numerous islands (photograph above). These were once the tops of hills that are now separated from the mainland by marine flooding. Cliffs, with rocky ledges or platforms at their base, occur along many of the coasts. Two types of platform can be recognised: subtidal and supratidal.

Subtidal platforms occur where shorelines plunge steeply into the sea in localities exposed to powerful ocean waves. Initially, waves undercut the steep hill slopes, which then collapse to form a cliff. As this retreats, a narrow platform (Figure A) develops just below the high tide level.

Wave action continues to undercut the cliff and remove the loosened material, generating a broader platform (B). This stage has been reached at Ping Chau. As the platform becomes wider, waves lose energy and become less erosive as they cross the ever-larger platform. Eventually, the waves are too weak to undercut the cliffs, and sand begins to accumulate (C). Scree (loose fallen material) collects at the base of the cliff, further protecting it from undercutting. The cliffs then gradually become gentler as a result of stream erosion and slope collapse.

The islands of the northeast, from Tiu Tang Lung. Two large bays, Crooked Harbour and Double Haven, are surrounded by islands. These are former hills that were cut off from the mainland by a sea level rise that began about 18,000 years ago. The present shoreline was established about 6,000 years ago.

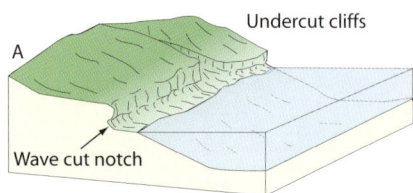

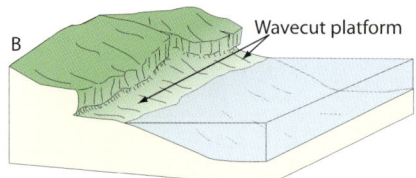

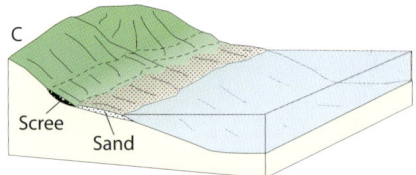

Subtidal platforms develop in areas subject to strong waves. Initially, a narrow shelf and small cliff form (A) as a result of undercutting by waves. With time, the shelf becomes broader (B). Waves lose energy as they cross broad platforms. Consequently, they begin to deposit sand (C). A well-developed platform on Ping Chau is shown below.

The Northeastern New Territories: Marine Inlets and Islands

Both Sha Tau Kok (Starling Inlet; above) and the Tolo Channel (below) were once river valleys. Their straight alignments are due to the ancient rivers preferentially carving out the fractured rocks along NE-SW-trending faults. The hills on the far side of Starling Inlet are in mainland China. The border passes through the town of Sha Tau Kok, on the far shore. Plover Cove Reservoir (below) lies to the north (left) of the Tolo Channel with Ma On Shan town in the lower right. The island at the end of Tolo Channel is Chek Chau (Port Island).

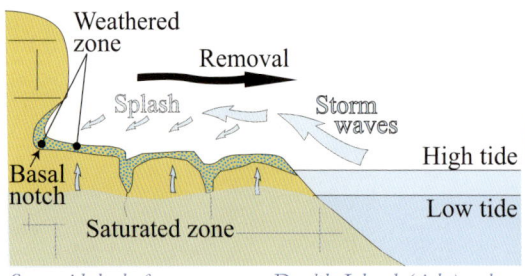

Supratidal platforms occur on Double Island (right), where they are formed by salt weathering in the splash zone above the high tide (figure). Some sea water may also seep through the rocks. The loosened rock is then removed by storm waves, gradually eroding the cliff backwards.

Supratidal platforms are flat shelves, several metres wide, backed by small cliffs, commonly with a basal notch (figure above). They are situated immediately above high tide level, and formed in tuffs susceptible to salt weathering. This process involves sea water, which enters cracks in the rock. As the water evaporates, salts crystallise. The crystals grow, exerting a pressure that forces adjacent particles apart, and which opens cracks wider, thus weakening the rock. This is an example of physical weathering, which can cause a rock to disintegrate grain by grain. The sea water is supplied by splash from waves, even though the platform lies above the high tide level. Once loosened by salt weathering, the material is removed by storm waves. If the process occurs on small islets then sombrero islands (middle right photographs) may develop. These posses a tiny central hill surrounded by a supratidal platform. Sombrero islands are only present in this part of Hong Kong.

Several islets in Double Haven show how sombrero islands evolve. They begin to develop when erosion forms a flat surface around a small central hill marked by a cliff (upper photographs). As the platform expands (right), the hill is reduced in size and is ultimately eroded away, producing a flat island.

The northeastern region also contains a fine example of a tombolo (right), a ridge of sand that joins an island to the mainland. The term was originally derived from an Italian name for a burial mound.

Ma Shi Chau, viewed from Yim Tin Tsai (northern Tolo Harbour). The narrow sandy ridge linking the island is called a tombolo.

Human Impacts

The Eastern Closed Frontier Area

The Closed Frontier Area was set up in 1951 in an attempt to prevent illegal immigration into Hong Kong. The eastern Closed Frontier Area originally passed over a ridge in northeastern Hong Kong that extends from Wo Keng Shan, through Robin's Nest to Hung Fa Chai, and included all of the waters of Starling Inlet (Sha Tau Kok) (right photograph). No entry was allowed into the area without a permit. The region was fenced with coils of barbed wire. Later, this internal border was shifted several kilometres northwards, passing just south of Lin Ma Hang and over Hung Fa Chai. In 2006, the government announced that after 55 years of isolation about 70% of the zone (2,000 hectares) would be opened for development, with some areas to be given ecological protection.

The eastern Closed Frontier Area was saved from development by its restricted status for more than 50 years. The region contains both lowland and mountain habitats of high ecological value and a diverse wildlife that includes: Malayan porcupine, wild boar, Chinese ferret badger, Indian muntjac, yellow-bellied weasel, crab-eating mongoose, Anderson's stream snake, and mountain wolf snake. Consequently, proposals have been made to designate parts of this area a country park, with priority for protection being given to the secondary forests and lowland streams. It has also been suggested that green corridors be established between the Wutongshan National Forest Park, Shenzhen and Robin's Nest, Hong Kong, in order to help cross-border wildlife movements.

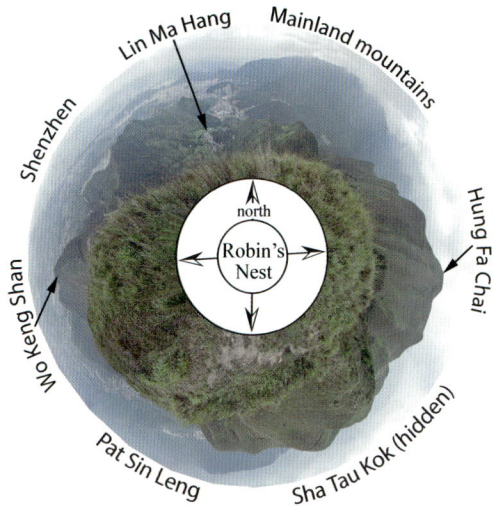

The 360° panorama above shows the view from the summit of Robin's Nest in northeastern Hong Kong. The Closed Frontier Area extends over the northern half of the view to the base of the mainland mountains.

The photographs show the formidable border fences that snake through the Sham Chun Valley, separating Robin's Nest from Wutongshan. The modern highway, in the background above, lies within mainland China and contrasts markedly with the undeveloped Hong Kong side of the border.

The Lin Ma Hang Lead Mine

Lin Ma Hang lead mine is situated in the Closed Frontier Area, about 2 km from the village of Lin Ma Hang (map, p. 69). Mining has left scars on the landscape in the form of tunnels, excavations, and waste dumps. The rocks are tuffs altered by metamorphism (IN11, p. 51), with the mineral resources being found in lense-shaped veins. These extend for 2 km and are a few millimetres to several metres wide. Some veins are vertical, but elsewhere they are gently inclined. There were two periods of vein formation. The first produced milky-coloured quartz (IN21, p.100), whereas the second phase generated a fine-grained quartz, which carried the metallic minerals that have been mined. These yielded: lead, zinc and copper.

The economics of mining was the driving force behind the human impacts on the landscape. Galena (IN21, p. 100) was the main source of lead and also carried silver of between 300 and 500 g per tonne. Ore was assayed at 10–12% lead and about 50–100 g per tonne silver. Lead concentrates were shipped to smelters in Europe, notably the United Kingdom and the Netherlands. At the time of closure in 1958, about 60% of the reserves had been mined and sold. Waste tips in the area are dominated by rock and quartz, the main gangue mineral (the valueless waste).

Lead was first discovered in the 1860s. The lowest workings of the mine, called the Portuguese Workings, date from the nineteenth century, whereas the main vein was discovered in 1915 by Chinese miners. Adits (tunnels) were driven into the hillside at three levels. These followed the ore-bearing veins: Level 0 was located at the top, Level 1 was driven 80 ft (about 24 m) below that and Level 3 was excavated 60 ft (18 m) deeper still.

The Lin Ma Hang mines once provided a livelihood for scores of lead workers who dug tunnels hundreds of metres into the northern side of Robin's Nest mountain. The adits, long abandoned, now serve as a home for large colonies of bats.

A mining company was initially formed in 1917 and in June 1925; a 75-year mining lease was taken out by Morrison Brown Yung, who operated the mine until his death in 1932. Following a brief period of limited operation by a Chinese company, Hong Kong Mines Ltd. acquired the lease and property in March 1937. The company had just been taken over (in January) by Nielsen & Co. Inc., an American firm based in the Philippines. During 1938, the peak period of production, the mine employed 350 people underground and 150 surface workers, and ore delivery was increased from 150 to 225 tonnes a day. With the outbreak of World War II, work was suspended in 1940. Small-scale mining continued during the Japanese occupation from 1941, but there was considerable damage done to the adits, partly because the Japanese had removed ore from rock pillars used to support the roof.

The entrance to one of the main mining chambers features two large pillars left to support the roof. These sometimes contained lead and, at other locations, were removed by the Japanese during World War II, causing roof collapse.

The mine was reopened in October 1951 and was worked by two successive contractors until 1954. The designation of the Closed Frontier Area in the 1950s restricted access to ten vehicles a day, and the contractors were required to build a new road into Lin Ma Hang Village in 1953. Eleven miners and 11 other staff were employed. Working methods were dangerous and the government ordered operations to be suspended. Work recommenced in September 1954, but labour disputes, typhoon damage, and falling lead prices led to the closure of the mine on 30 June 1958. The mining lease finally lapsed on 13 April 1962.

Mine tailings lie scattered across parts of the hills at Lin Ma Hang (below). These contain a variety of minerals, such as lead-grey galena (top) and gold-coloured pyrite (above).

Hill Fires

About 5% of the land area of Hong Kong is burnt annually. Most fires result from carelessness, many being caused when offerings are set alight at grave sweeping ceremonies. Hill fires probably first affected the natural vegetation on a large scale in the twelfth century with the arrival of the Han, who cleared land for agriculture, commonly by fire. The fire season coincides with the dry period (October to April), with about 850 hill fires occurring each year. Figures specifically for the Country Parks (1990–91) show a total of 184 fires burned 2,244 hectares with a loss of 89,400 trees. Fires deplete the topsoil and remove protective vegetation, leaving soil prone to desiccation and erosion. Without destruction by fires, undisturbed grassland would probably be replaced by scrubland in about 10 years, followed by woodland being established within 50 years.

A December 2004 fire ravaged the southern (left) scarp of Pat Sin Leng. Fortunately, the fire stopped at the summit ridge, saving the vegetation and wildlife on the northern side from considerable damage.

Plover Cove Reservoir

During the early 1960s, the growing demand for potable water, coupled with several years of low rainfall, led to water shortages and severe rationing. So in November 1966, six Hakka villages were evacuated to make way for a new reservoir. Opened in 1968, Plover Cove Dam impounded a marine bay in Tolo Harbour. The main dam is 2 km long and 44 m high, with subsidiary dams between islands. An extensive existing natural catchment was augmented by the construction of 30 km of tunnels that brought water from adjacent watersheds. The dam was raised in 1973 to increase the storage capacity from 170 million m^3 to 230 million m^3.

Plover Cove Reservoir occupies a former bay off Tolo Harbour. Marine mud had to be removed, and the area desalinated before fresh water was introduced. The photograph shows the reservoir in the foreground viewed southwards from Tai Tung. The main dam lies in the extreme right of the photograph. This mountain ridge forms part of a rugged circular trail that includes the Bluff Head Ridge in the middle distance. The Tolo Channel and hills of Sai Kung Country Park form the background.

Marine Parks

The marine resources and environments of Hong Kong have been under significant pressure from over-fishing for decades. Trawling and the use of dynamite and cyanide have contributed to a decline in the fish stocks and damage to coral reefs. Four marine parks and one marine reserve have been set up to protect valuable habitats from further destruction. Two of these lie within the northeastern New Territories: Yan Chau Tong and Tung Ping Chau.

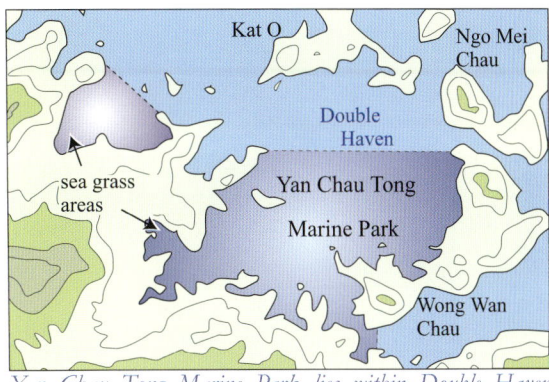
Yan Chau Tong Marine Park lies within Double Haven and between a ring of islands and the Hong Kong mainland.

Yan Chau Tong Marine Park occupies the southern half of Double Haven, a circular bay between mainland Hong Kong and the islands of Kat O, Ngo Mei Chau (Crescent Island), and Wong Wan Chau (Double Island). The park was designated on 5 July 1996, with a total area of about 680 hectares. It protects mangroves, a rich diversity of corals, dominated by *Platygyra* and *Favia*, and several patches of seagrass, which are found on sand and mud.

Tung Ping Chau Marine Park (270 hectares) was designated on 16 November 2001. It encircles the island of Ping Chau in northeastern Mirs Bay. Over 30 species of hard corals have been recorded, in addition to 130 reef-associated fish, over 100 marine invertebrates, and more than 40 species of algae. Ecologically valuable sea grasses protect sandy seabeds from erosion.

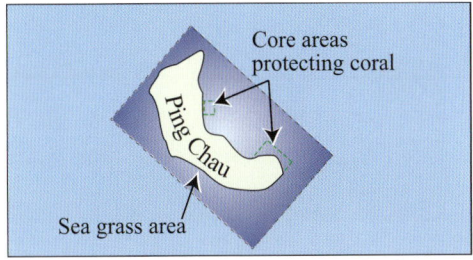
Ping Chau was the fourth marine park to be designated and protects both coral and sea grass communities.

Coral reefs at Tung Ping Chau have survived damage from fishing activities, and are now protected within a recently designated marine park.

The Northeastern New Territories: Human Impacts

Villages and the Landscape

Villagers have been modifying habitats in the region for several centuries. In some places, such as Wu Kau Tang (map, p. 67), a thriving community remains. In contrast, villages such as Lai Chi Wo (p. 67 and below) struggle to survive, having lost many inhabitants. Others (e.g. Kop Tong, Mui Tsz Lam; p. 67) are abandoned.

The people who built these communities had a profound impact on landscape. Hills and lowlands were terraced for rice. Natural materials were used to build homes. Communities were laid out according to a Feng Shui plan that sought environmental harmony (IN15, p. 83).

The Hakka village of Lai Chi Wo shows several of these Feng Shui principles, and their basis in common sense. The settlement lies between two ridges on the southern (sunny) slopes of Pan Pui Teng. A biologically rich woodland uphill from the village protects it from cold winds and landslips. The village stands above a floodplain (allowing it to remain dry after rain), which is used for rice cultivation.

The Feng Shui and other woodlands near Wu Kau Tang harbour a rich biodiversity, with 164 tree species having been recorded. Also, over 30 dragonflies and more than 50 butterflies have been observed in the area. In 1997, it was proposed that the main stream be designated a Site of Special Scientific Interest (SSSI), with 85% of the location being given the status of a Conservation Area. Nevertheless, proposals have been made to use the region for purposes such as a herbal theme park.

Kop Tong is located high on a hillside to the south of Lai Chi Wo. The village has been totally abandoned, as have several others in the Plover Cove Country Park.

The village of Lai Chi Wo is shown in the map and the photograph above (viewed from the south). Note the location of the woodland that rises up the hillside above the village in a classic Feng Shui plan. Ridges to the north and south form an armchair shape, which is also a major Feng Shui motif.

IN15 Feng Shui and Landscape

Feng Shui means *wind* and *water* and is based on a belief in earth forces that control health, prosperity, and luck. The concept involves a balance between Yin and Yang. Yin represents female, negative and passive aspects of the world, whereas Yang is correlated with male, positive and active characteristics. Feng Shui beliefs originated over 3,000 years ago when people noted that they fared better in certain settings. For example, it was observed that south-facing hills provided protection from strong northerly winds.

The Feng Shui woodland at Kop Tong (above) preserves a stand of old trees and provides a wildlife refuge.

Feng Shui is also based on *Chi*—a force that pervades landscapes and life itself. The Chinese character for Chi has two meanings, one cosmic and one human. Cosmic Chi involves air, gas, weather, and force. Human Chi involves breath, aura, manner, and energy. Places that are notably influenced by Chi are also the most pleasant and habitable. Chi is sometimes viewed as spiralling around within the earth, occasionally rising toward the ground, and at others places descending inwards. When Chi reaches the surface, mountains are created and, where Chi descends, deserts develop.

Mountains and water are important in Feng Shui, so the shape of the landscape must be considered before building a house. Dragons are the most common mountain symbols that protect many villages. Different parts of mountains represent various segments of a dragon's body. For example, a series of knolls along a ridge may be seen as the dragon's vertebrae, with secondary ridges on either side representing the limbs. Mountain streams are considered to be dragon veins. Auspicious locations include hill summits—the meeting place of heaven and earth, and the southeastern side of mountains, where there is plenty of warmth from the sun. Inauspicious locations include flat plains with no rivers, and ridges, which represent the tail of a dragon (that might thrash). Having a home below an overhang is bad because, like a dragon's upper jaw, it might catastrophically drop.

The optimum setting is the protective "armchair hill", sometimes called the *mother-embracing child*. This site is backed by a high *tortoise* mountain, with a lower *green dragon* hill to the left and a *tiger* mountain to the right (slightly lower than the dragon hill). Ideally, a small hill (the *vermillion bird*) should rise from the flat lands just below the village on the opposite side to the tortoise mountain. A stream should also run through the site with houses being placed part way up the tortoise mountain.

Feng Shui woods had a practical value: protecting a village from landslips, soil erosion, and winds, as well as supplying fire wood. Often they assumed a mysterious status that protected them from further change and development. Today, Feng Shui groves contain old and diverse trees, forming valuable habitats for local wildlife.

The Western New Territories

Parts of the western New Territories demonstrate the most extreme examples of gully erosion in Hong Kong. Other areas show how deliberate human intervention can bring about a recovery in "natural" landscapes. The region includes the Castle Peak Range, the Tuen Mun Valley, and the hills of the Tai Lam Country Park, as well as parts of the low-lying Shap Pat Heung and Shek Kong floodplains. Many small villages lie within the latter two areas. The largest town is Tuen Mun, which housed about 500,000 people in 2003.

There are three distinct mountain massifs. The Castle Peak Range is bordered to the west by Deep Bay and to the east by the broad Tuen Mun Valley. This mountain range includes a sharp ridge on its eastern flank that reaches a high point at Castle Peak (583 m). Most of the other hills consist of low, rounded, grassy summits that show the effects of severe erosion.

The foothills of the Castle Peak Range descend to the western coastline of Hong Kong at Deep Bay. The area is characterised by low rounded hills strewn with numerous boulders produced by the weathering of the underlying granite rocks.

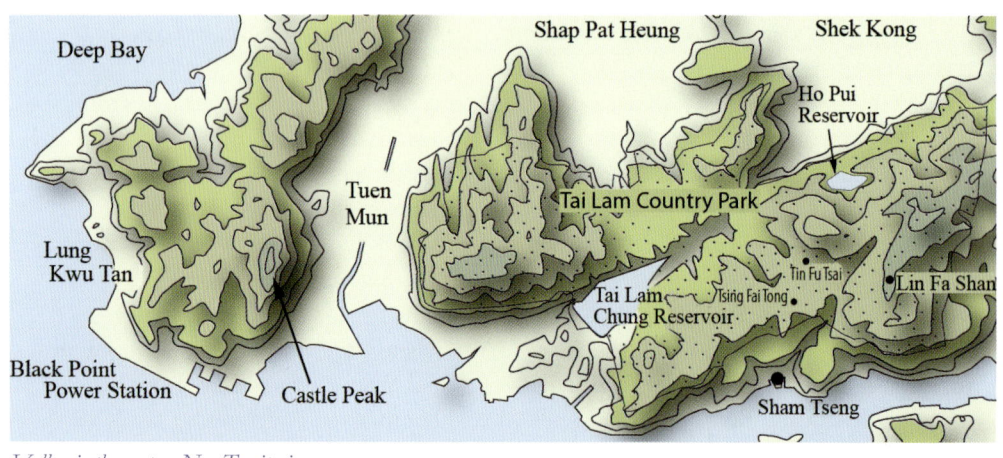

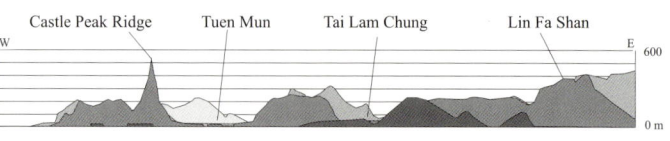

Valleys in the western New Territories follow a NNE-SSW or a NE-SW trend. Note the strong alignment of the southern Shek Kong Plain and the Tai Lam Chung Valley, which lie along a major fault line. The Tuen Mun Valley also follows a structural alignment. Most of the hills are rounded and relatively low, with the highest point being Castle Peak at 583 m. The composite profile shows the view from the south.

The two other upland areas lie to the east of Tuen Mun and are separated by the Tai Lam Chung Reservoir and its valley. Both sets of hills lie within the Tai Lam Country Park and were heavily eroded as a result of deforestation, particularly during World War II, when there was a fuel shortage.

Acacias, pines, and eucalyptus trees were planted, from 1952 onwards, to protect the catchment area of the Tai Lam Chung Reservoir from serious erosion that was denuding the hills and fouling the reservoir (IN20, p. 98). As a result, the Tai Lam Country Park and adjacent areas are now covered by extensive forests that give the scenery a different appearance to that of the eroded Castle Peak Range. In recent decades, these woodlands have been infested by microscopic worms (nematodes) transmitted by beetles; many trees have died.

Forests cover the once severely eroded Tai Lam Country Park. Today, the area provides habitats for barking deer, pangolins, Chinese leopard cats, and numerous bird species.

Proximity to extensive forests offers an escape from urban life, even on cloudy days, that few other cities can match. This forest is part of the Tai Long Nature Trail, at the southern end of the Pat Heung Valley, near Yuen Long.

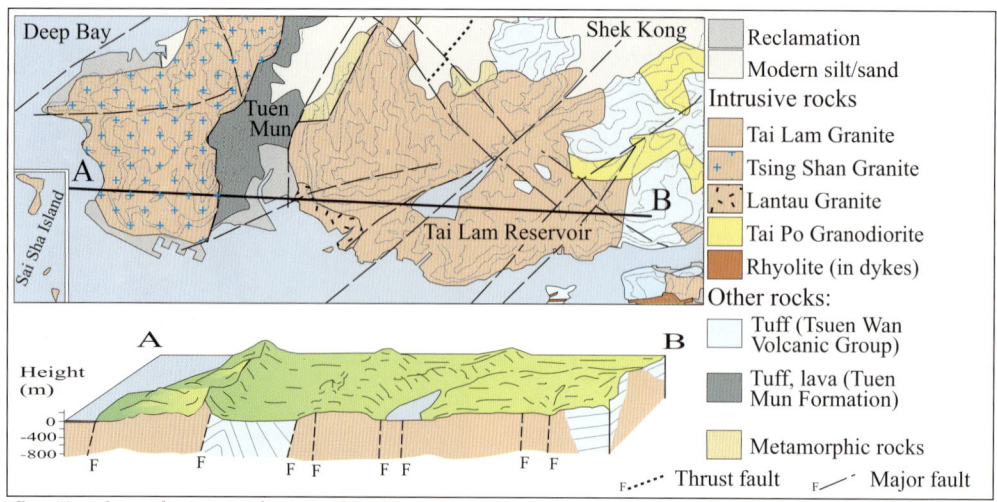

Granitic rocks are dominant in the western New Territories and on the Sai Sha Islands. Volcanic rocks constitute the remainder, except for a small belt of metamorphic rocks near Tuen Mun. Note the close relationship between valleys and fault lines.

Magma Chambers and Eruptions

Granites lie beneath most of the western New Territories. These rocks have been weathered and eroded to produce rounded hills and mountains (p. 21–23). Originally, they formed in large underground intrusions called plutons (IN04, p. 22). In some cases, the hot magma pushed closer to the surface to form magma chambers that fed overlying volcanoes (figure above right). The magma in these chambers cooled slowly over hundreds of thousands of years, eventually solidifying to form granitic rocks in the Hong Kong region.

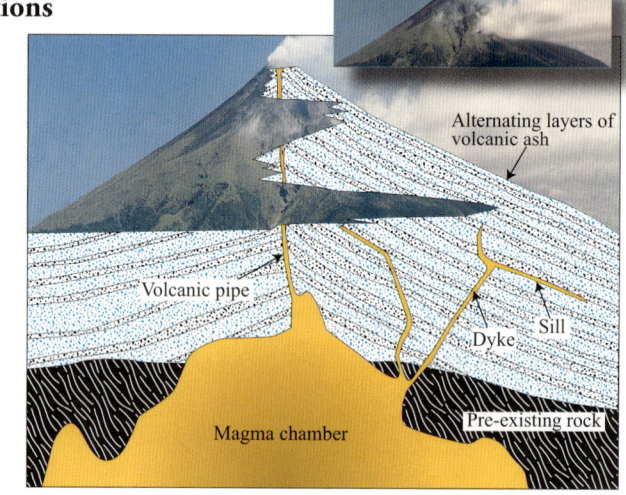

Two kinds of granitic rock are common locally (adjacent figure; map above), giving rise to subtle variations in the landscape. These are granite and granodiorite. The former tends to decay in a manner that produces white to yellow colours on exposed weathered surfaces. In contrast, granodiorite generates duller, more reddish colours. These differences are also visible along footpaths. Granodiorite, for example, breaks down to produce

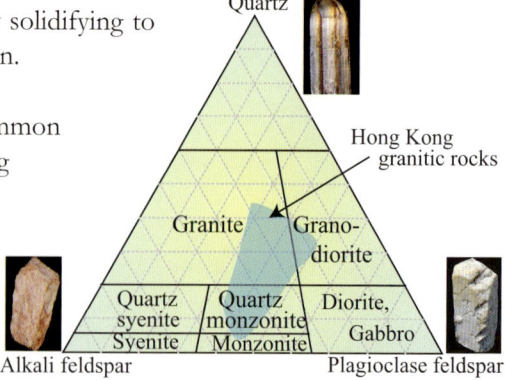

Granitic rocks are distinguished by the content of quartz and two kinds of feldspar. Hong Kong rocks contain a little of each, with granite and granodiorite being dominant. Dark-coloured minerals also occur but are not shown in the diagram above.

clay-rich footpaths that can be very slippery in wet conditions. Soils developed on granite tend to be sandier with abundant quartz grains. These differences are due to the two rock types containing different proportions of the minerals: quartz (glassy grey), alkali feldspar (grey to pink; rich in sodium), and plagioclase feldspar (grey to white; rich in calcium). Granodiorite includes a larger percentage of plagioclase feldspar and other dark-coloured minerals (not shown in triangular diagram, p. 86), such as hornblende and biotite.

The magmas in this region were emplaced during the first of four phases of granitic intrusion that took place in a back-arc geological setting (adjacent figure; IN23, p. 105). During this first phase of activity, several magma bodies formed at slightly different times and places. The granodiorites, for example, are restricted to the east of the western New Territories and are 164 million years old. The various granites are more widespread and their magmas were intruded later. A small area of Lantau Granite (161 million years old) occurs to the southeast of Tuen Mun. The Tsing Shan and Tai Lam Granites formed 159 million years ago in two separate, elliptically-shaped plutons.

Volcanic rocks are also present in the western New Territories and were erupted as either loose ash (termed tuff when converted to rock) or lava. In Hong Kong, volcanic eruptions took place during five distinct time intervals. The first (180 million years ago) predates the granitic intrusions noted above and gave rise to tuffs, lavas, and sediments of the Tuen Mun Formation (map, p. 86). Four later periods of volcanism occurred at about the same times as the four phases of granitic intrusion (adjacent figure). The Tsuen Wan Volcanic Group (in the north and east) accumulated 164 million

The Tai Lam Granite contains large rectangular crystals of pinkish feldspar and abundant, grey, greasy-looking quartz, as well as minor quantities of black biotite.

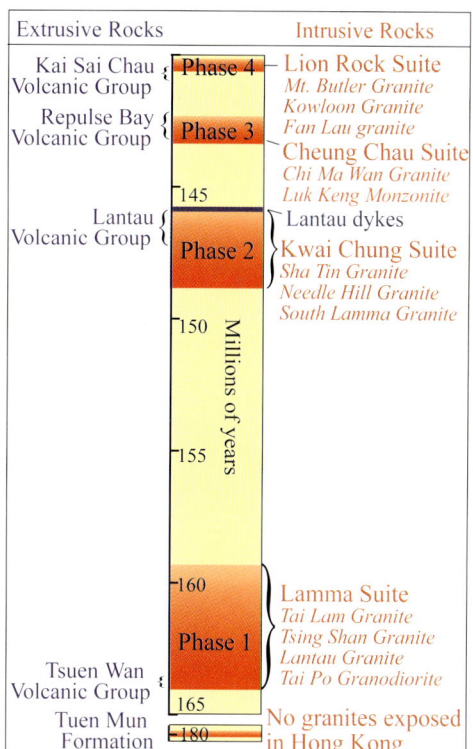

Extrusive rocks in Hong Kong formed from volcanoes during five time intervals. The first of these occurred 180 million years ago. The four later periods of volcanism coincided with four phases of granite intrusion, separated by episodes with no igneous activity. The volcanic rocks have been assigned to Groups and the granitic rocks have been placed together in Suites.

years ago at about the same time as the Lamma Suite of granitic rocks (figure, p. 87).

The mineral quartz occurs as veins in several parts of the western New Territories. Quartz usually crystallises from hot fluids that squeeze through cracks in pre-existing rock during the late stages of granitic intrusions. The quartz veins may be small, or large as in the adjacent photograph. Being relatively hard, they commonly form low ridges that cross the countryside.

The photo above shows a white quartz vein following a fault that cuts through the Tai Lam Granite. The quartz crystallised from hot fluids that flowed along the fault. Where there were large spaces, quartz was able to grow without interference from adjacent rocks, forming large crystals as shown in the inset.

High Ground, Low Ground

The surface of the earth is not a flat plain, but is irregular. This is because the shape of the ground reflects a battle between two forces that vary from place to place. One force builds up the surface, generating mountains. The energy for this is derived from the internal heat of planet earth, which is produced by radioactive decay. This heat drives plate tectonic processes (IN16), which can push the ground upwards or downwards, depending on the precise processes that are taking place at a particular location.

IN16 Plate Tectonics

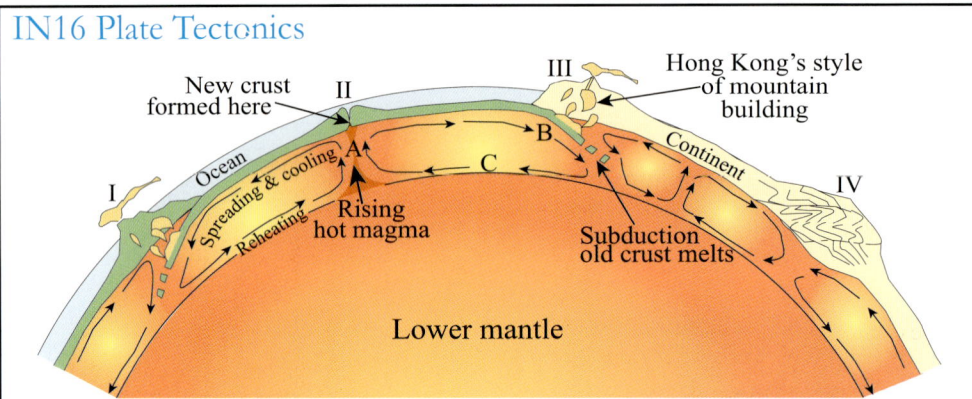

Plate tectonic processes depend on convection cells in the upper mantle. These cells are driven by heat from radioactive decay. Hot magma rising at A spreads towards B, dragging the overlying crust with it as it slowly cools. At B the cell descends causing the crust to sink and melt. The flow continues towards C, reheating the material, and eventually rising again at A. Uplift, and the creation of new land, occurs at the boundaries between crustal plates. For example, at I melting of the sinking crust produces magma that rises to form a new volcanic island (e.g. Philippines). At II the crust is lifted by upwelling magma, creating new crust, and a mid-ocean ridge develops (e.g. the mid-Atlantic ridge). At III an ocean plate descends below a continent, melts, and adds new material to the continent. This was the situation that gave rise to the volcanic mountains of Hong Kong 164–140 million years ago. At IV two continents collide, causing folding and thickening of the crust (e.g. the Himalayas).

The Western New Territories: High Ground, Low Ground

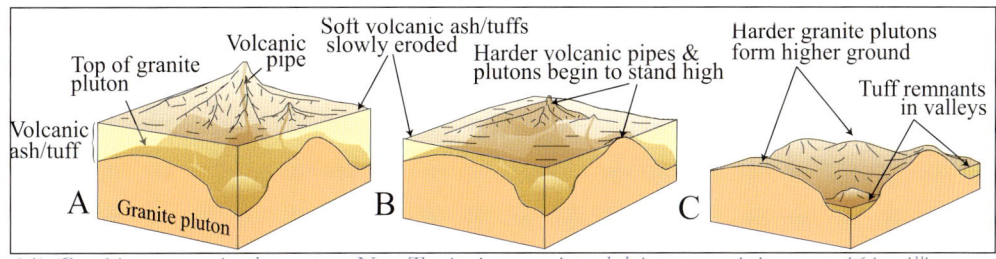

(A) Granitic magmas in the western New Territories were intruded into pre-existing crust 164 million years ago. Some of this reached the surface via volcanic pipes to form volcanoes. (B) The magma cooled, solidified, and erosion removed overlying material. (C) Once the ground was reduced to the level of the granites, differential erosion became important. The resistant granite was slowly reduced in height so that it formed hills. The pre-existing tuffs, metamorphic rocks, and sediments were generally weaker and eroded faster to form low ground.

The other force, erosion, reduces the height of mountains. Erosion processes are driven by solar energy and gravity. Heat from the sun evaporates water, which rises to form clouds and then rain. Subsequently, gravity causes the water (or ice) to move downhill in rivers (or glaciers), eroding the surface.

Granites in the western New Territories solidified from magmas that were intruded near a plate boundary (III in IN16, p. 88). When the plate tectonic processes ceased, erosion became the dominant force. Erosion has a profound effect over long periods of time. For example, if 1 mm (the thickness of a fingernail) is eroded away each year, then in 1,000 years the ground would be lowered by 1 m. Given a million years, 1 km will be removed. It is 160 million years since the granites were intruded—enough time to remove 160 km of rock. In reality, erosion is slower than this, but there would still be sufficient time for erosion to bring to the surface the granites that originally formed a few kilometres deep (figure above). Once the granites were exposed, another facet of landscape evolution, differential erosion (IN07, p. 37), resulted in the granites mostly forming mountains.

Differential erosion takes place if erosion proceeds at different rates at nearby locations. This occurs where particles and cements within a rock vary or where there are different rocks next to each other. These processes can operate on a variety of scales to produce hills and valleys or smaller scale features. For example, quartz veins (adjacent photographs) commonly

Erosion causes hard quartz to stand proud of softer rocks on a variety of scales. For example, the lower photograph shows 1–3-cm-thick quartz veins forming small ridges. The upper photograph is of a 4-m-wide quartz vein.

The Western New Territories: Gullies and Badlands

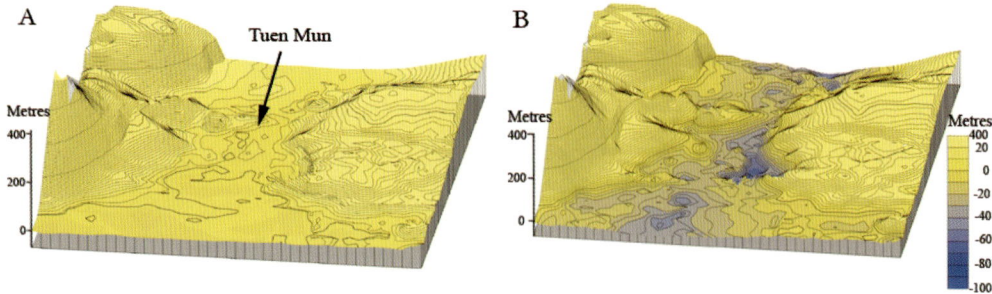

erode more slowly than the rocks they intrude. Consequently, they usually stand, fin-like, above the adjacent materials.

Similarly, granitic rocks erode more rapidly than tuffs, and height differences develop between areas underlain by these two lithologies. Furthermore, rocks are also weakened by faulting so that they erode faster. For example, a major fault runs through the Tuen Mun area and is responsible for the location and orientation of the valley (figures above).

Gullies and Badlands

Gullies are deep (greater than 1 m), steep-sided, erosion channels cut into weathered rocks, or loose colluvium (IN17, p. 91) on hillsides. They tend to occur where there is a lack of vegetation and where there are heavy downpours. Areas with dense concentrations of gullies are called badlands.

Severe gullying was probably triggered in the western New Territories by several episodes of deforestation. Gullies are especially common in the Castle Peak Range, where they are clearly

Block A shows the topography of the Tuen Mun Valley, with its flat floor composed of loose river deposits that rest on highly weathered rock. At the base of these soft materials is the solid bedrock. The surface of this bedrock is highly irregular (Block B, with the loose sediment removed). Note the aligned deep depressions (blue colour), which mark locations where weathering has decomposed the rock to great depth along a major NE-SW-trending fault. Over long periods of time, this weathered material is removed faster by erosion then the less-decayed materials from the adjacent unfaulted rocks. The differences in weathering and erosion rates are responsible for the formation of the Tuen Mun Valley—another example of differential erosion, this one controlled by faulting.

The Castle Peak Range (below) is characterised by treeless hills that have been heavily eroded. This desolate landscape is the best place in Hong Kong to see the development of rills and gullies. Hiking in the area provides wide open vistas, but the region is especially hot in the summer due to reflections from the light-coloured, weathered granites that are exposed along the main trails.

IN17 Colluvium

Colluvium is an accumulation of soil and rock fragments that are deposited on slopes by a variety of processes that reflect the influence of gravity. Examples include movement down a slope of small particles following rain splash, as well as continuous downhill creep, sliding, and flowing. Colluvium is usually unsorted and loose, containing particles that range from clay and silt to massive boulders.

Coarse boulder colluvium, D'Aguilar Peninsula.

Generally, colluvium is structureless, but some colluvial deposits exhibit a weak layering, which may indicate several phases of deposition. In some cases, differences in the weathering of the rock fragments and the finer-grained matrix distinguishes the layers.

Medium-coarse colluvium, Ngau Chi Wan, Kowloon.

Deposits of colluvium, more than 2 m thick, cover approximately 15% of the area of Hong Kong. Colluvium forms sheets or fans on the slopes of the higher peaks such as Tsing Shan, Tai Mo Shan, and Lantau Peak and forms extensive mantles over the Kowloon footslopes and in the Mid-Levels on Hong Kong Island. The largest deposits are up to 30 m thick. On the steeper slopes, colluvial sheets are commonly cut into by streams.

Volcanic colluvium extends over about 9% of Hong Kong, occurring as bouldery deposits in valleys or as fan-shaped accumulations. Numerous landslide scars are common in volcanic colluvium, evidence of general instability. Granite colluvium covers only 3% of Hong Kong, usually as boulder fans, particularly on the Kowloon foothills and northern Hong Kong Island. Sedimentary colluvium is restricted to 1% of the area, and is sandy or gravelly. Mixed colluvium covers about 2% of the surface.

Fine colluvium, Mid-Levels, Hong Kong Island.

Old Colluvium

Old colluvium formed many years ago, but is not actively accumulating today. It occurs on some hillcrests below large hill masses or as a capping to ridgelines, indicating a long period of stream erosion since it was formed. Clear examples are found in a lobe-shaped deposit at Wu Kwai Sha, and on the foothills of Fei Ngo Shan, where the colluvium is considerably weathered. Gullying and internal erosion are distinctive features of old colluvium, and disappearing streams are common. Old colluvium may also extend offshore, probably reflecting accumulation at a time when sea levels were lower.

Gully erosion is widespread in many parts of the Castle Peak Range. In places, they may be several tens of metres deep.

visible. They are also widespread in the Tai Lam Country Park, although they are now largely hidden as a result of extensive reforestation programmes. These erosional landforms are closely associated with granitic terrains because the underlying rocks are especially prone to chemical decay (IN18), which produces loose sandy particles that can easily be removed by flowing water. In many areas, the weathered rock is several tens of metres thick, which allows very deep gullies to develop.

Gully Formation

Gullies may evolve in several ways. In Hong Kong, the two most important processes involve stream incision and upstream erosion by spring sapping. The former occurs when heavy rain falls on steep, weathered slopes. Water initially flows downhill as a very thin layer (sheetwash) spread evenly across the hillside (figure, p. 93). Surface irregularities cause the water to concentrate into discrete threads of

IN18 Granite Weathering

Hong Kong's hot humid climate promotes chemical weathering. This involves several processes that bring about the decay of rock. The most important in granites involves the conversion of feldspars (which form up to 80% of granitic rocks) to a white clay mineral called kaolin. This is soft, and easily eroded. Quartz, the second most common mineral in granites, is much more resistant to chemical breakdown and remains unchanged. As granite decomposes, it gradually changes colour and becomes weaker, allowing it to erode more easily.

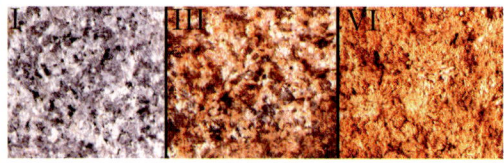

Six weathering grades are recognised by geologists in Hong Kong. Grade I is fresh unaltered rock (left). Grade II is hard, but stained near cracks. Grade III (centre) is moderately decomposed. Grade IV can be broken by hand into smaller pieces. Grade V is completely discoloured and breaks into individual grains. Grade VI (right) is a residual soil with total loss of original rock textures and which easily falls apart.

Rills (photographs above) develop when water flow is concentrated. If bare loose soil is exposed, particles can be easily removed and streams cut rapidly downward into the underlying materials. In extreme cases, deep, steep-sided gullies may grow (adjacent photograph). Note the flat top of this gully, which once was a steep military road with a thin protective seal. After it was abandoned, the seal broke down and erosion proceeded rapidly on the steep gradient.

flow, which become more concentrated down slope. Rills (photographs above) begin to develop when the speed of the water is sufficient to detach and remove loose particles. As these grains are carried away, rills gradually become wider and deeper, evolving into larger features called gullies. A small waterfall may develop over the steep upstream face during heavy rains. As the water pounds the floor, it erodes the base, causing the upper slope to be undercut. Eventually, collapse occurs, and the gully will extend further upstream.

Gullies may also be generated by spring sapping. Water flows below the surface, as well as above, and may emerge from the ground at a spring. In Hong Kong, springs are usually associated with soil pipes (adjacent figure). These are small tunnels in the soil along which water flows quickly. As the water exits from a pipe, it removes particles. Eventually, this undermines the soil layer above and collapse occurs. The resulting gully then gradually extends upslope.

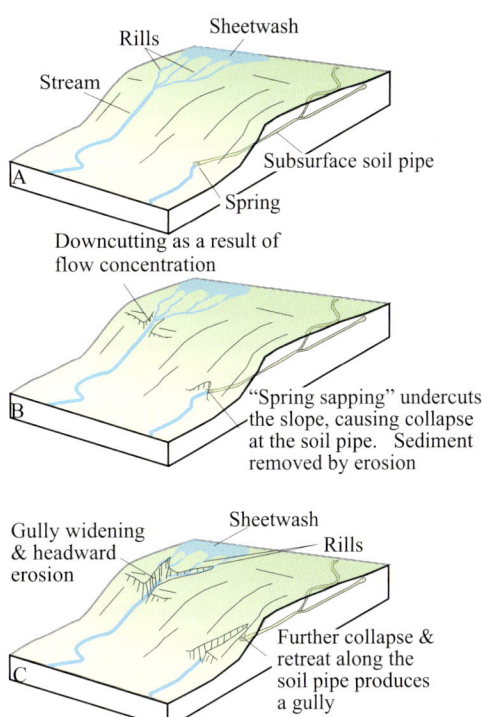

Gullies may form by either stream incision or spring sapping associated with collapse of materials overlying a soil pipe. In both cases, slow headward erosion extends the gully uphill.

Mountain Streams

The bleak appearance of the eroded hills of the Castle Peak Range is partially relieved by the presence of several pleasant streams. The largest is the Tsing Tai River, which flows from south to north through the centre of the region, entering Deep Bay at Nim Wan by Tai Shui Hang Village. This is a typical mountain river with an irregular long profile that is interrupted by several waterfalls (adjacent photographs).

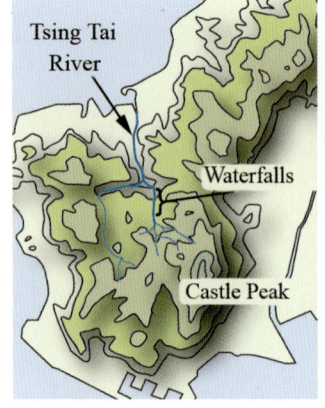

Mountain streams have considerable energy by virtue of their steep gradients. Some of this energy is used to overcome friction along the very irregular long profile, which slows the water. Remaining energy is used to erode the channel floor down to base level (IN19, p. 95) and to carry sediments of various sizes. The energy of a stream is related to flow speed, which varies considerably, even over short distances. Consequently, in mountain streams, it is common to find areas of erosion close to sites where boulders have been deposited. In some cases, finer-grained sediments, such as sand, may accumulate as well. Sand, boulders, or other particle sizes, of course, can be deposited only if they are made available by erosion within the stream's drainage area; thus, source rocks also exert an influence on the nature of a stream.

Waterfalls usually form where bands of hard rock resist erosion, although they can also develop as a result of base level changes (IN19, p. 95). They are sites characterised by rapid variations in stream energy. The drop over a fall imparts energy to the water, which allows it to carve out a depression at the bottom called a plunge pool. In the left photograph note the presence of both sand and boulders. These are areas of deposition during floods, with the sand indicating where the stream energy is lowest.

IN19 Base Level and Streams

As streams erode, they widen and deepen their valleys, creating smooth, long profiles. However, there are several factors that can prevent this development. A major control is the river's base level. This is the level below which the stream cannot erode. In the upper diagram of the figure to the right, the base level is the same as sea level (also called absolute base level). At this height, the water slows down and erosion stops. As time passes the profile will be lowered from 1 to 2, but it cannot cut below the absolute base level because streams cannot flow uphill to the sea.

Sea level is not fixed—it can rise or fall. The lower diagram shows what happens if absolute base level falls. A waterfall may form at the former coastline, then slowly retreat upstream as the long profile is lowered from 2 to 4. During this retreat, the waterfall gradually will be lowered until it forms rapids and then ultimately disappears.

A stream may also have temporary or local base levels along its course. One example is a lake. The stream cannot cut below the lake surface, which will therefore control erosion patterns upstream until the water body is drained by incision at its outlet. A hard rock layer can also act as a local base level, forming a waterfall, until it too has been eroded away.

This waterfall in Tai Lam Country Park shows how irregular the long profiles of mountain streams can be.

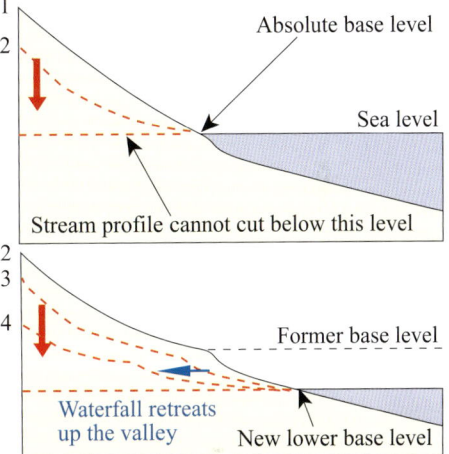

The sand pictured above accumulated where a stream entered the Tai Lam Reservoir at a time when its surface stood just above the sand and constituted a local base level (figure below). The sand was then incised by the stream because the reservoir level (or base level) had fallen.

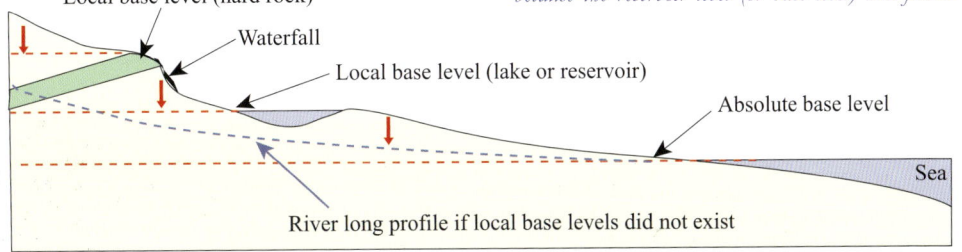

Human Impacts

Archaeological evidence suggests that human occupation near Tuen Mun dates back to the New Stone Age (about 10,000–5,000 years ago), but it is only during the last few hundred years that the impact of people on the landscape has been significant. Tuen Mun, for example, became a port for Sino-western trade towards the end of the Tang Dynasty (618–907 AD). Troops were stationed at a sentry post, which became the origin of the name of Tuen Mun (garrisoned entrance). For a short period during the Ming Dynasty, the Portuguese occupied Tuen Mun before being driven out by Chinese naval forces. Military links continued through the colonial era, with much of the Castle Peak mountains being designated as an army firing range. The area is still closed to the public during army field-training exercises.

For much of its settled history, the Tuen Mun Valley was dominated by farming and fishing, with salt production near the coast. This changed dramatically with the development of the New Town. Early construction began in 1968, with large scale projects taking place from the 1970s. About 275 hectares of land were reclaimed from the sea, which once extended much further inland (map above). During the development phase, the population of the town increased from 20,000 to 530,000.

To the north of Tuen Mun and in the Pat Heung Valley, the lowlands are occupied by innumerable small villages. Several ancient trails crossed the mountains and connected the different districts. The Yuen Tsuen ancient trail, for example, runs for 13.5 km across the Tai Lam Country Park, linking the Pat Heung Valley

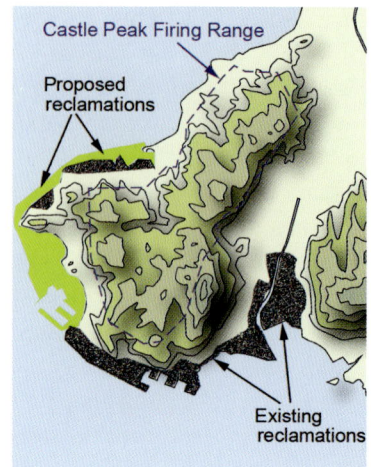

Much of Tuen Mun is built on land reclaimed from the former Castle Peak Bay. Other reclamations, such as for the Black Point power station, extend around the southern end of the Castle Peak Range. Proposed further reclamations extend even further to the west.

The panorama below shows the New Town of Tuen Mun and its valley from the Castle Peak Range. The hills beyond and to the right are part of Tai Lam Country Park. Note the hazy conditions. These are typical for much of the year, especially during the winter months when monsoon winds blow from the north bringing pollutants from the rapidly expanding industries of the Pearl River Delta. In general, western Hong Kong, including Tung Chung on Lantau, experience the worst atmospheric pollution in Hong Kong.

with Tsuen Wan. This trail once provided the only means for villagers in Shap Pat Heung to transport their produce to markets in Tsuen Wan and to obtain non-agricultural supplies and utensils in exchange. The forest section, Nam Hang Pai, was paved with natural stones about 3–400 years ago. Similar routes also helped other isolated mountain villages to thrive.

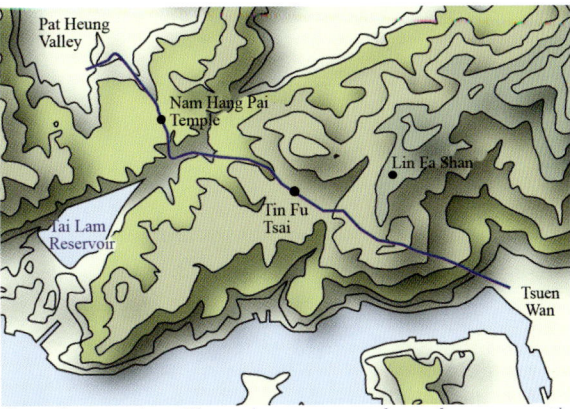

Over the centuries, villagers have constructed roadways across the Tai Lam Country Park. The example shown in blue is the Yuen Tsuen Trail, which links the Pat Heung Valley with Tsuen Wan.

It was noted earlier that widespread erosion has occurred as a result of deforestation. However, in recent decades, successful attempts have been made to reverse the destruction. Afforestation (IN20, p. 98) was initially carried out to protect the water catchment area of the Tai Lam Chung Reservoir and prevent it from slowly being choked with sediment. The building of a dam was originally suggested in the 1930s, but it was not until 20 years later, after World War II and following further forest destruction, that tree planting actually commenced. Soil conditions were poor, and, in some places, forests were planted directly on weathered rock with no organic soil layer. Problems occurred due to outbreaks of fire,

Small temples, such as this one on the Nam Hang Pai Trail, were built at major crossing points over the hills of Tai Lam Country Park. They are testimony to the cultural importance of the routes.

The appearance of the landscape varies dramatically in the western New Territories. The Castle Peak Range (left), having lost its protective vegetation cover, was subjected to extensive erosion, particularly along the ridge lines. Tai Lam Country Park (above) experienced a similar level of erosion, but reforestation has stabilised the environment and concealed most of the old scars. The panorama above also shows Tsuen Wan in the foreground, with Tsing Yi Island across the Rambler's Channel (right). Hong Kong Island can be seen in the background.

IN20 Plantation Trees

Until recently, the Tai Lam Country Park was heavily eroded following past deforestation. To prevent further erosion, trees have been planted extensively since 1946. In 1983, for example, 0.8 million trees were planted in the country parks of Hong Kong. The main desirable characteristics of these trees include: fast growth, low nutrient and water requirements, and an ability to withstand fire. A few examples of trees commonly used are illustrated.

*The paperbark tree (*Melaleuca leucadendron*) grows up to 24 m tall and is native to Australia. It is easily distinguished by its peeling bark and can be found in both wet and dry sites.*

*Slash pine (*Pinus ellioti*) was introduced from North America and grows to 30 m tall. It differs from local red pine in having a more regular conical shape, longer needles (18–25 cm), and being more resistant to fire and pests. It thrives on almost all soils and is commonly the first species to be planted.*

*Ear-leaved acacia (*Acacia auriculiformis*) is widely used in afforestation because of a symbiotic relationship between its roots and bacteria that are able to introduce nitrogen into nutrient-poor soils.*

Acacia mangium (left) is an introduced evergreen tree that grows rapidly, reaching 3 m in only 3 years. Keteleeria fortunei (centre) is an evergreen that grows to 27 m tall. The leaves are arranged in two ranks, and it has winged seeds in large cones. Eucalyptus trees (right) were introduced from Australia because they can survive in dry conditions and are fire resistant.

but, eventually, mature stands of Taiwan acacia, ear-leaved acacia, big-leaved acacia, keteleeria, brush box, red pine, slash pine, and eucalyptus were established.

In the 1950s, the area around Lin Fa Shan (Lotus Flower Hill) (map, p. 97), in Tai Lam Country Park, experienced a mining boom after the black mineral wolframite (IN21, p. 100) was discovered. The area became known as Black Gold Hill. Wolframite contains tungsten, a valuable metallic element that can be used to strengthen other metals. Seeking their fortune, several thousand stakeholders dug up to 1,000 exploratory adits (small tunnels) in an area of less than 1 km². At the time, the region was dominated by grasslands and scarred with waste from mining activities (photograph below). Today the region has largely recovered, with woodlands covering most of the area and concealing the scarred landscape.

Partially hidden entrances to derelict adits are common on Lin Fa Shan. They can be dangerous but are reminders of the work of thousands of miners in bygone days.

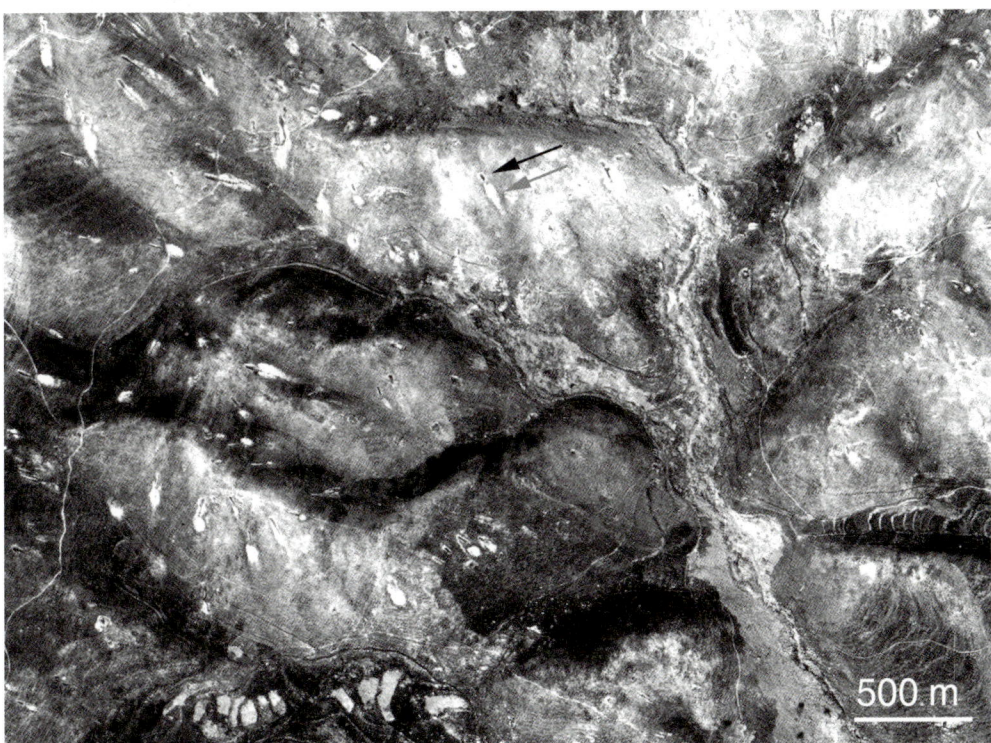

This January 1963 aerial photograph was originally taken from a height of 3,900 feet (1,188 m), and shows Lin Fa Shan, which was once covered with mining adits. The entrances show up as small black marks (black arrow). Lighter patches on the downslope side (grey arrow) indicate debris thrown out from the adit as mining proceeded. Note the high density of the mines in the west (left) Today, the area is heavily forested, and the adit entrances are difficult to find.

IN21 Hong Kong Minerals

There are many thousands of minerals, but only a few are economic, and even fewer have been mined in Hong Kong. Four locally mined examples are shown on the right.

Similarly, only a few minerals make up the bulk of most rocks. These include quartz, which is a mostly milky-white, greasy-looking mineral (B), though it sometimes forms clear, six-sided crystals (A). Calcite is also white but much softer (unlike quartz, it can be scratched with a penknife). Large calcite crystals form rhomb shapes, as in photograph (C). The most abundant minerals are feldspars (there are several types). These are usually opaque and white, light grey, or slightly pink (D). Two dark-coloured minerals are also common in Hong Kong, especially in granite. These are biotite (E), which splits into flat, paper-like masses, and hornblende, which is black and tends to have a prismatic shape (F).

Galena (PbS, lead sulphide) was mined at Lin Ma Hang (p. 78–79). It has a metallic grey colour, and a blocky appearance. The mineral is soft and can be scratched, with difficulty, by a fingernail. The mineral is especially heavy with a density of 7.4 (7.4 × heavier than water). It is a major source of lead.

Wolframite ((Fe,Mn)WO$_4$, ferro-manganese tungstate) was extracted from the slopes of Needle Hill and Lin Fa Shan. This is a brownish-black mineral with prismatic crystals. It has a red-brown streak (the colour of the powder when the mineral is scraped onto a surface). It can be scratched with a knife (but not a fingernail) and is dense (7.3). The mineral is a major source of the metal tungsten.

Molybdenite (MoS$_2$, molybdenum sulphide) occurs at Needle Hill but was never a commercial resource. It has a lead-grey colour and is moderately heavy (5.5). The mineral has a metallic sheen, a green-grey streak, and is very soft. Molybdenum is used as an alloy in the manufacture of stainless steel. The hexagonal crystals in the picture are quartz.

Sphalerite (ZnFeS, zinc iron sulphide) occurs at Lin Fa Shan, Needle Hill, Devil's Peak, and Lin Ma Hang but was not mined economically. It may be brown, yellow, green, or black, and has a density of about 4. It possesses a greasy appearance and has a brown-white streak. The mineral is a source of zinc.

THE CENTRAL NEW TERRITORIES

The landscape of the central New Territories, more than most regions, reflects Hong Kong's violent eruptive past. The region is dominated by three mountain peaks, underlain by volcanic or intrusive rocks. The highest peak is Tai Mo Shan (Big Hat Mountain; 957 m), and the others are Grassy Hill (647 m) and Needle Hill (532 m). The Sha Tin Valley is also underlain by igneous rocks, but the scenery there has been altered significantly by people.

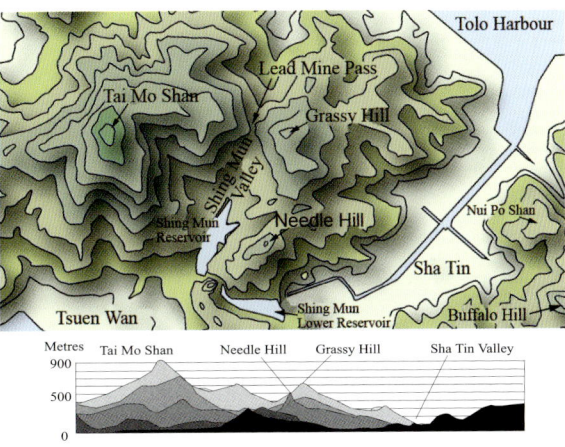

Tai Mo Shan is the highest peak in Hong Kong and, together with Grassy Hill and Needle Hill, forms part of a large circular massif. The Sha Tin Valley, to the east, follows a major NE-SW-trending fault that extends beyond this area to include Tolo Channel and the Tai Po Pass. The west-east oriented profiles are viewed from the south.

Streams radiate from the Tai Mo Shan massif in all directions, with several being interrupted by waterfalls (IN22, p. 102), including some of the highest in Hong Kong. At the centre of the highland block lies the NE-SW-trending Shing Mun Valley and the Lead Mine Pass. Both have interesting histories. In the mid-seventeenth century, Ming Dynasty rebel soldiers occupied the southwestern base of Needle Hill and built a fort (Shing) and gate (Mun) to control the area. They raided local villages until their surrender to the Ching Dynasty army in 1656. The fort was built close to Pineapple Dam. Lead from galena (PbS) was once mined at Lead Mine Pass and was used for debasing coins. Unfortunately, the deposits were of poor quality, and the mine ceased operating in 1898.

Tolo Harbour, to the northeast, was once the centre of a pearl fishery. The Sha Tin Valley was originally a tidal estuary surrounded by sandy (Sha) fields (Tin). Reclamation and development over the last three decades have converted it into a conurbation of over 630,000 people. The region also includes parts of Tsuen Wan (270,000 residents in 2003), which was built on land reclaimed from Gin Drinker's Bay.

The high mountain ranges are characterised by extensive montane grasslands and offer broad panoramas of the surrounding region.

IN22 The Ng Tung Chai Waterfalls

Middle (Horse-tail) Falls

Scatter Falls

The Main (Long) Falls

Bottom (Well) Falls

The four main Ng Tung Chai waterfalls lie in Tai Mo Shan Country Park and are approached from the village of Ng Tung Chai. Bottom Falls (also called the Well Falls) is divided into lower (7 m) and upper (17 m) portions by a small plunge pool. Middle Falls (Horse-tail Falls) is 17 m high. The largest of the group is Main Falls, which has one of the longest drops (35 m) in the territory. Scatter Falls lies on the highest part of the stream, and is divided into two sections (10 and 7 m high). A landslip resulted in the official closure of the path between Main and Scatter Falls, but this has been re-established by enthusiasts who continued to walk the route. A fifth falls (not shown), Maiden Falls, lies above the four shown in the photographs and consists of a series of smaller drops, partly obscured by vegetation.

The Central New Territories

The northern slopes of Tai Mo Shan rise from the Lam Tsuen and Shek Kong valleys and are clad with forests, scrub, and high montane grasslands. Several trails, some very rough, follow the ridge lines towards the summit. The smaller, but distinctive, peak of Kwun Yam can be seen rising behind the ridge line to the right.

Grassy Hill lies in the middle of the central New Territories. Here it is viewed from the MacLehose Trail as it ascends Tai Mo Shan. The extensive forests to the right of the peak are part of the Shing Mun Country Park.

The major urban centre of Sha Tin lies within a long fault-controlled valley that is surrounded by high mountains. The hazy outline of Tai Mo Shan can be seen in the background, with Grassy Hill interrupting its right-hand flank. Needle Hill cuts the left-hand slope. This photograph was taken from Buffalo Hill and looks northwestwards.

An Era of Violent Eruptions

The entire Tai Mo Shan massif and Sha Tin Valley are underlain by igneous (volcanic and intrusive) rocks. This area exemplifies the difference between the age of a landscape and the age of the underlying rocks. Although the rocks were formed more than 160 million years ago, the mountains and valleys were carved much more recently—in just the last few million years. Today, these landforms are still being reshaped by erosion. Thus, a thorough appreciation of the scenery of any area requires an understanding of the geological past. The present mountains and valleys would not exist without the distinctive igneous rocks that formed many millions of years ago.

The first eruptions in Hong Kong (180 million years ago) occurred in a volcanic arc above a subduction zone (IN23, p. 105). This activity produced lavas and tuffs (Tuen Mun Formation, p. 87) from cone-shaped volcanoes. As time passed, the nature of the volcanism changed. The subduction zone migrated to the southeast, and the area that is now Hong Kong was then located in a back-arc setting (IN23).

The back-arc eruptions probably took place from fissures and collapsed volcanoes called calderas (IN40, p. 136), during four distinct phases of activity (figure, p. 107). The first phase produced tuffs of the Tsuen Wan Volcanic Group (TWVG). These rocks were erupted 164 million years ago and now form the highest mountains in the central New Territories. Four subdivisions have been recognised within the TWVG (adjacent figure). The oldest is the Yim

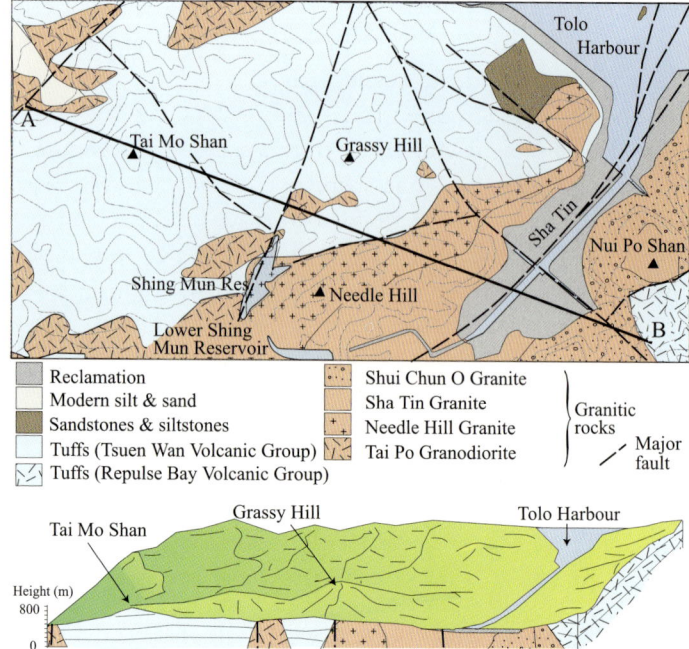

Tuffs form the highest mountains in the region, including Tai Mo Shan. Lower ground around Sha Tin is underlain by granitic rocks.

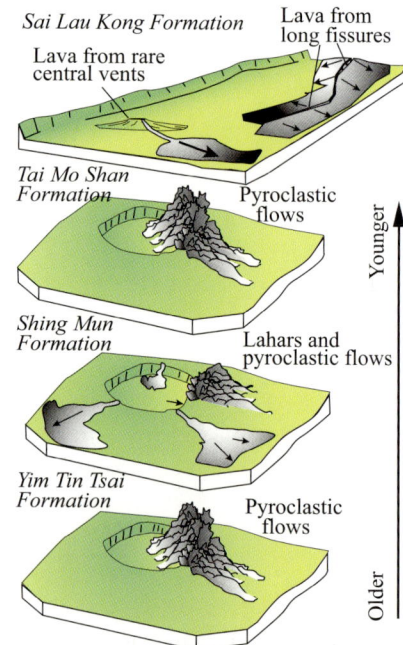

Volcanic environments changed during the accumulation of the TWVG. The landscape was dominated by caldera volcanoes that varied in their activity. Periods of intense volcanism were separated by relative quiescence, during which erosion dominated and lahars were formed.

IN23 Migrating Volcanoes

Two kinds of volcanic setting can be recognised near continent-ocean plate boundaries (figure below; IN16, p. 88). Volcanic arcs form close to subduction zones. They are fed by magma rising from a melting descending ocean plate. These magmas tend to produce tall, cone-shaped volcanoes. Back-arc basins form further away from the plate boundary, and are fed by magma circulating in a convection cell (corner flow) between the two plates. Drag against the overlying continental plate creates tension, faulting, and thinning of the crust, which allows magma to ascend from the mantle. In Hong Kong, this back-arc setting was associated with caldera and fissure eruptions.

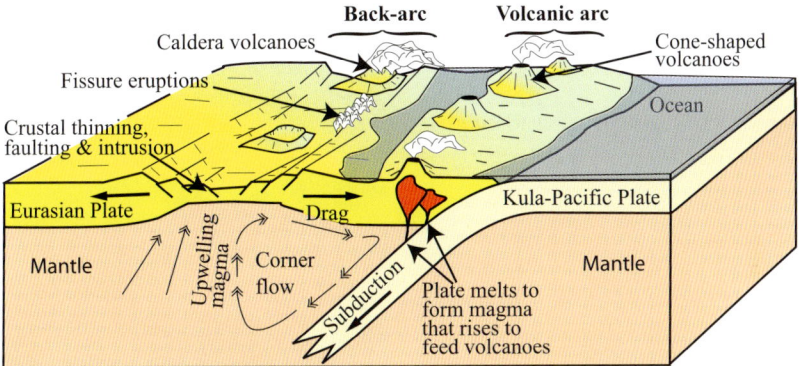

The earliest volcanic rocks (Tuen Mun Formation) in the territory were erupted in a volcanic arc setting. However, the style of eruptions changed because the plate boundary was migrating southeastwards across the area that is now Hong Kong (adjacent figure). As a result, the territory came under the influence of back-arc volcanism, with fissure and caldera eruptions (IN40, p. 136) occurring during four distinct phases.

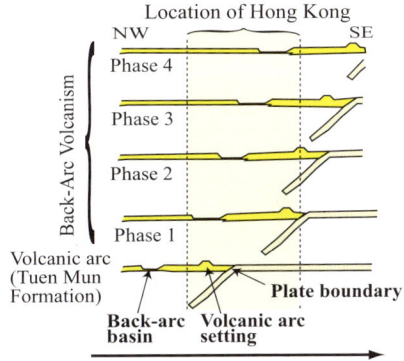

Back-arc volcanism took place during four igneous phases (map right). During Phase 1, calderas were produced in northwestern Hong Kong. By Phase 2, the boundary had shifted southeastwards, and eruptions took place along what is now the Sha Tin Valley. Continuing shifts in the plate boundary resulted in volcanism drifting to the southeast. The last volcanic activity ceased 140 million years ago.

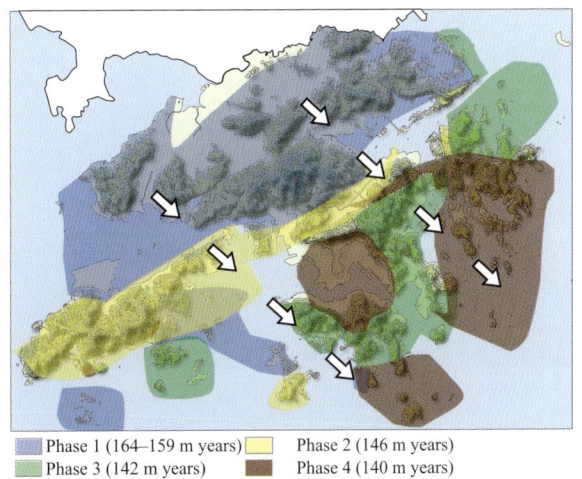

Phase 1 (164–159 m years) Phase 2 (146 m years)
Phase 3 (142 m years) Phase 4 (140 m years)

Tin Tsai Formation, which is dominated by coarse-grained tuffs. These eruptions produced pyroclastic flows, which consist of mixtures of ash, volcanic gases, and air. Being more dense than the atmosphere, they flowed rapidly down the sides of the calderas. A period of reduced activity followed, and the volcanoes began to erode. Rain mixed with loose ash to form dense liquid mixtures (lahars) that flowed downhill. These ancient lahar deposits now constitute a major part of the Shing Mun Formation (block diagrams, p. 104).

The Tai Mo Shan Formation is dominated by tuffs. Here, two volcanic particles have been flattened. Originally rounder, they were squeezed, while still hot, under the weight of overlying ash.

The third and fourth divisions of the TWVG are the Tai Mo Shan Formation and the Sai Lau Kong Formation (p. 104). The former includes coarse-grained tuffs that signify a return to intense eruptions. In contrast, the latter formation is dominated by lavas that erupted from fissures. However, these rocks are restricted to the northeast New Territories and are not present within this region.

Crags on Kwun Yam Shan (below) indicate the presence of hard and resistant rocks that were formed in a volcanic pipe (a conduit linking a magma chamber to a volcano). These rocks are rhyolites (IN04, p. 22) (adjacent photographs) that cut through the Shing Mun Formation. They have been broken and altered by hot fluids rich in silica, replacing rock fragments and filling voids with quartz (IN21, p. 100).

These distinctive rocks (rhyolitic hyaloclastites), on Kwun Yam Shan, were fragmented by contact between hot rocks and water. Siliceous fluids deposited patches of white quartz in a series of parallel layers.

Kadoorie Farm viewed from the northern slopes of Tai Mo Shan. Note the distinctive peak of Kwun Yam to the left, which consists of a hard and resistant altered rhyolite. The remainder of the valley is underlain by tuffs of the Shing Mun Formation. The latter rocks extend to the first summit of Tai To Yan in the middle background. The second more distant peak on the same ridge is underlain by the Tai Mo Shan Formation.

The Central New Territories: An Era of Violent Eruptions

The Shui Chuen O Granite forms the peak of Nui Po Shan (Turret Hill), viewed in this photograph from Buffalo Hill. Note the typical rounded form of a granitic landscape.

Granite crops out locally on Nui Po Shan and forms distinctive rock features. Note the sides of this pillar and the horizontal crack at its base. These are joints that have controlled the shape. The steep slope in the background is made of rhyolite.

Eastern and southern parts of the region are underlain by granitic rocks that originated in magma chambers (IN04, p. 22). These granites are younger than the tuffs of the TWVG (adjacent figure). The oldest is the Tai Po Granodiorite. This forms a ring-like intrusion that was injected into the Shing Mun Formation. The ring structure is centred on Tai Mo Shan with areas of granodiorite encircling the mountain (map, p. 104).

Other granitic rocks were intruded in more recent phases of activity (adjacent figure). For example, the Needle Hill Granite and Sha Tin Granite were intruded during Phase 2, whereas the Shui Chuen O Granite formed several million years later during Phase 3.

The granites developed in areas that lay to the southeast of earlier igneous activity and were associated with a back-arc geological setting (IN23, p. 105), where prevailing tensional forces pulled the rocks apart. This created a series of NE-trending faults that allowed magma to escape through the crust. The most significant example is the Lai Chi Kok-Tolo Channel

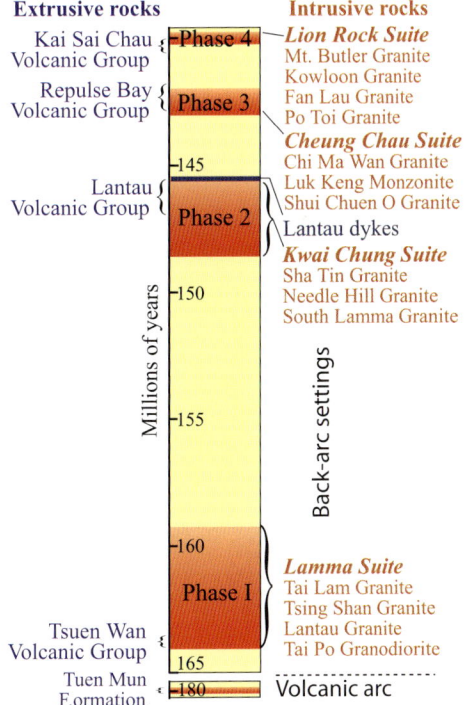

The earliest eruptions in Hong Kong were associated with a volcanic arc (IN23, p. 105). Subsequently, there were a further four distinct phases of igneous activity that were associated with a back-arc setting (IN23). These later eruptions produced granitic rocks at depth and volcanic materials at the surface.

The Central New Territories: An Era of Violent Eruptions

Fault Zone. This 60-km-long line of weakness consists of a 750-m-wide belt of crushed rocks and discrete fractures that lies parallel to other faults along the coast of southern China. The fault zone exerts a strong influence on the topography because the crushed rocks are easily weathered and removed by erosion. Consequently, a series of valleys and passes have developed along this alignment, including the Sha Tin Valley, Tolo Channel, and Tai Po Pass. The fault zone may have been active during the intrusion of the Sha Tin Granite and Needle Hill Granite (148–146 million years ago). However, it is inactive today, as are most faults in Hong Kong. Consequently, there is only a moderate risk of damaging earthquakes (IN24).

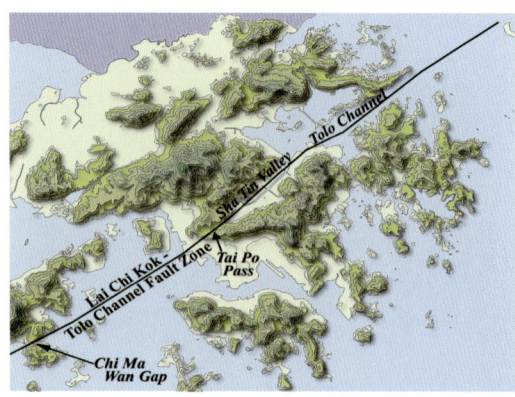

The Lai Chi Kok-Tolo Channel Fault Zone extends across Hong Kong along a northeasterly alignment. This structural line of weakness has eroded faster than adjacent areas to produce a series of valleys, channels, and mountain passes.

IN24 Earthquake Risks in Hong Kong

Hong Kong has never been the epicentre of a major earthquake in recorded history. However, 25 quakes with a Richter magnitude greater than 6 have occurred within 650 km of Hong Kong in the last 1,000 years. Only two events took place within 250 km. In the twentieth century, there were at least 72 of magnitude 5–6. In Hong Kong, earthquakes are few, not noticed, and mostly less than 4 on the Richter scale. Significant earthquakes close to Hong Kong have included a 6.0 magnitude event at Hong Hai Bay on 15 May 1991; a 5.5 tremor near Macau on 12 August 1905; and a 5.75 incident near the Dangan Islands on 23 June 1874. A 1918 earthquake at Nanao resulted in the strongest tremors ever felt in Hong Kong, but even this only caused minor damage to buildings.

Earthquakes can be measured by either magnitude or intensity. Magnitude indicates the amount of energy released and is measured by the open-ended Richter scale (map). The intensity at any particular place indicates the strength of the motion produced there by an earthquake. This is measured by the 12-point Mercalli scale, which notes the effects of a tremor on people and buildings. The destructiveness of a quake is related to its magnitude and the local geology. For example, loose sediments tend to produce more damage. Buildings in Hong Kong are able to withstand tremors of VII on the Mercalli scale—the maximum rating expected locally.

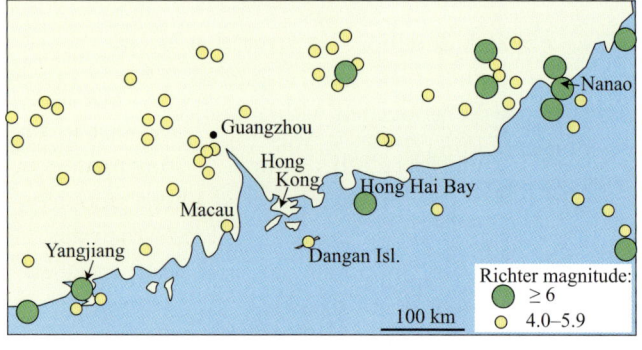

Human Impacts

Human impacts have taken a variety of forms over recent centuries. Before the arrival of people, the central New Territories were wooded. Settlement brought with it clearances for village construction, agriculture, and charcoal production. Rice was grown on the lowlands, with upland varieties planted to heights of at least 450 m. Tea was cultivated in the highlands, leaving a unique imprint (IN25).

These changes were not all visually undesirable, as they increased scenic diversity, adding new environments to the old. Landscapes comprise more than just hills and valleys and human imprints. They also include the vegetation that clads the slopes and the wildlife. Enjoyment of the scenery is enhanced by the sight of an eagle overhead or the sound of birds in trees. Animals also interact with the landscape. Termites, for example, influence soil formation (IN39, p. 134). Wildlife is an indicator of a healthy landscape. Although the last tiger was shot as late as 1947 in Sha Tin, the territory still supports a varied wildlife (IN26, p. 110).

Due to the steep slopes in Hong Kong, the boundaries between towns and countryside are sharp. The photographs show Sha Tin (upper) and Tai Po clearly defined by surrounding mountains.

IN25 Tea Terraces

The occurrence of tea plantations on the upper slopes of Tai Mo Shan were reported in the *Sun On Gazetteer* in 1688. Tea was grown on obliquely inclined stone terraces, which are best seen today after a hill fire. Several authorities have suggested that they are probably much older and predate Hakka settlement of the area in the seventeenth and eighteenth centuries. Construction of the many terraces would have involved a huge amount of labour. The tea (*Camellia sinensis*) produced was referred to as shan cha, or mountain tea, and was particularly astringent and favoured by the elderly. Tea obtained from wild plants was also known as wan mo, or cloud mist tea.

Tea terraces on the higher slopes of Tai Mo Shan are obliquely inclined, probably as a drainage-enhancement measure. The examples shown in the photograph are on the northern slopes above Kadoorie Farm.

IN26 Animal Biodiversity

The great variety of available landscape habitats in Hong Kong is reflected in the rich biodiversity of the territory. Mammals include barking deer, pangolin, civet cats, wild boar, and macaque monkeys. Birds are even more varied. A few examples of the local wildlife are illustrated.

1. Spotted Dove (*Steptopelia chinensis*). Widespread nesting in bushes and trees a short distance above the ground.
2. White-bellied Sea Eagle. Seen in coastal areas. Feeds on fish, sea snakes, and young water birds.
3. Crested Goshawk. Lives in forests and woodlands. Feeds on birds, small rodents, and lizards.
4. Black-eared Kite. Widespread along coasts, rivers, and urban areas. The most common bird of prey. Feeds or scavenges on birds, mammals, fish, and insects.
5. Koel. Common and widespread throughout year. Noticed in spring due to its distinctive call, "Ko-el", repeated several times.
6. Leopard Cat. Nocturnal, feeds on birds, small mammals and frogs. Mainly in woodlands. Endangered and listed under CITES.
7. Burmese Python. The largest (up to 6 m) snake in Hong Kong. Kills by constriction.
8. Wild Boar. Nocturnal and solitary. Omnivorous, eating roots, leaves, insects, and small mammals. Sleeps in a 3–4 m deep burrow.
9. Rhesus Macaque. Although originally native, the ancestors of these monkeys were released in the 1920s. They are diurnal and will eat virtually anything.

The Central New Territories: Human Impacts

The Shing Mun Reservoir

The Shing Mun Reservoir, which lies at the centre of the Tai Mo Shan hills, is also known as the Jubilee Reservoir in honour of the Silver Jubilee of King George V. It was impounded to deal with water shortages (IN27). Construction took place between 1933 and 1937, with the building of three dams. The largest is 85 m high and the reservoir has a capacity of 3,000 million gallons. Pineapple Dam blocks the western side of the reservoir and was named after pineapples grown there by local Hakka farmers. Eight villages, with 855 people, were evacuated when the dams were built. The former inhabitants had owned 73 hectares of land and had forestry rights to 40% of the Shing Mun Country Park (IN28, p. 112). Most were resettled in Kam Tin. A deep, steep-sided gorge lies below the main dam (photograph above, p. 113). There have

The photograph above, from Kam Shan Country Park (IN28, p. 112), shows the Shing Mun Reservoir in a deep valley between Tai Mo Shan and Needle Hill, with its gullied and eroded slopes. Dams were constructed at several locations, including the entrance to the Shing Mun Gorge, which drains towards Sha Tin (right in photograph). The lower part of the gorge is hidden in the bottom right but is shown in the image below, where it has been dammed to form the Lower Shing Mun Reservoir, seen here during the dry season.

IN27 Reservoirs

Settlements on Hong Kong Island initially relied on streams for water, supplemented from 1851 by five wells. The first reservoir was opened in 1859 in the Pok Fu Lam Valley and was subsequently expanded in 1871. Tai Tam Reservoir was constructed in 1889 and included a 2.5-km tunnel to transfer water to Victoria. The scheme was expanded, and Wong Nei Chung was opened in 1899, with three further Tai Tam Reservoirs between 1904 and 1917, plus the Aberdeen Reservoirs in 1931 and 1932. The Kowloon Reservoir, in Kam Shan Country Park (IN28, p. 112), opened in 1910 and was followed by the Shek Lei Pui (1925), Reception (1926), and Byewash (1931) Reservoirs. The Shing Mun Reservoir opened in 1937. These were followed by the construction of the Shek Pik Dam (1963) on Lantau. Subsequently, two marine inlets, Plover Cove (1968) and High Island (1979), were dammed and filled with fresh water.

IN28 Country Parks

Hong Kong has a total area of only 1,092 km², but about three-quarters is countryside. The fact that this wealth of natural resources has been protected and survives today is due to earlier foresight in planning for the future, with the realisation in the 1960s that there was a need for conservation.

Work on protection commenced with the publication of the Talbot Report in April 1965, which was prepared for the International Commission on National Parks. Their report, "Conservation of the Hong Kong Countryside", led to the formation of a Provisional Council for the Use and Conservation of the Countryside in 1967. In June 1968, the Provisional Council recommended that eight specific regions be designated for recreation, that several areas be set aside as nature reserves, and that a permanent advisory body be established.

Under the guidance of Sir Murray MacLehose, who became Governor in 1971, the Country Parks Ordinance was enacted in August 1973. This sanctioned a Country Parks Authority and a Country Parks Board. Within 3 years the Country Parks Authority determined the boundaries for 21 Country Parks covering 40,000 hectares, or 40% of the land area of Hong Kong, and recommended an implementation programme between 1976 to 1981. The Country Parks are currently managed by the Agriculture, Fisheries and Conservation Department.

The popularity of the parks has been demonstrated by increasing numbers of visitors, with 12.2 million guests being recorded in 2005 alone. Today, there are 23 Country Parks. Fifteen Special Areas, such as Tai Po Kau and Tung Lung Fort, have also been designated, eleven of which lie inside the parks.

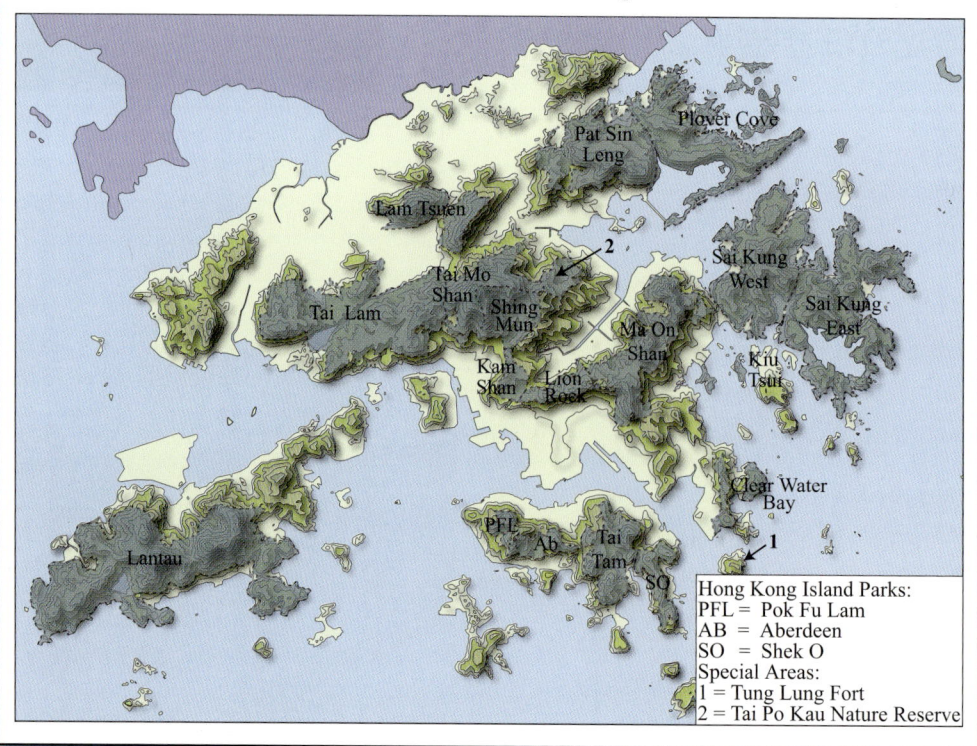

been several rockfalls on the valley walls, which were cut back during reservoir construction. For example, a major failure occurred along joints (IN09, p. 37) in June 1997, releasing large blocks and slabs. Above this collapse, on Needle Hill, there are many eroded slopes, the result of damage caused by mining (IN29, p. 114).

Population Pressures

Over the last 200 years, Hong Kong has experienced several waves of immigration that transformed the landscape. Major influxes resulted from the Tai Ping Rebellion (1850–1864), the Japanese invasion of Guangdong (October 1938), the Japanese surrender in 1945, and the Communist victory in 1948–49. The last influx led to unplanned, congested, and insanitary squatter settlements, mostly in Kowloon and on Hong Kong Island. The dangers were highlighted on Christmas Day 1953 when a fire destroyed the Shek Kip Mei squatter village, leaving over 50,000 people homeless.

The Public Works Department immediately erected two-storey emergency blocks, re-housing 35,000 of the homeless within 53 days. Subsequently, six- and seven-storey blocks were built. These were managed by the Resettlement Department, formed in 1954 to clear and re-house squatters. In 1972, the Governor (Sir Murray MacLehose) launched a 10-year Public Housing Programme, which was given further impetus by a renewed influx of immigrants from China at the end of the Cultural Revolution and by major fires during dry weather in 1981. These fires left 12,000 people homeless, over half (6,400) in the largest fire at Tai Hom Wor. In late 1981, there were 144,000 squatters and an additional 163,000 people on the public housing waiting list. Heavy rains in June 1982 caused disruptive landslides in many squatter areas, particularly at the Lam Tin village. Successive natural disasters led to an acceleration of the Temporary Housing Programme from 100,000 to 150,000 person-spaces a year.

The Shing Mun Gorge connects the Shing Mun upper and lower reservoirs and is the deepest in Hong Kong. Note the large landslide, on the right side of the gorge, which may have collapsed as a result of the side walls being deliberately cut back.

Densely crowded squatter communities remain in parts of the New Territories. Many were cleared in response to deadly fires in the 1980s. These photographs show squatter settlements at San Po Kong (Kowloon) in 1994 (above) and at Sham Tseng in 2000 (below).

IN29 Mining and Landscape Scars

Lead (from galena; IN21, p. 100) occurs together with zinc and copper in quartz veins in the central New Territories. Lead Mine Pass, for example, was named after small scale operations that occurred there in the nineteenth century. These diggings had little impact on the landscape, but a more significant resource, wolframite (IN21), lead to significant scarring. This latter mineral is a source of tungsten and was discovered in 1935 below Needle Hill by Mr. G. Hull, who was panning in a stream at lunchtime. The deposits occur in a series of parallel, NW–SE-trending quartz veins (map). Other minor, uneconomic resources include: molybdenum (from molybdenite), sulphur (pyrite), and fluorine (fluorite).

Mr. Hull obtained a license (later transferred to Marsman Hong Kong China Ltd.). Mining began in 1938 with three adits used to extract ore. The operations closed in 1941 after the Japanese invasion, but resumed between 1942 and 1945 under Japanese control, with two more adits developed. Tungsten prices increased between 1949 and 1951, which stimulated a boom in unlicensed surface excavation, with up to 5,000 workers at several sites. Severe erosion resulted, and government controls were introduced in 1954 to prevent further landscape damage. From 1951, the operations were run by the Hong Foo, and later the Yan Hing, mining companies. Mining ceased in 1967.

Adits are tunnels dug into a hillside. The top photograph shows examples on the eastern slopes of Needle Hill. These were small scale, as shown by the lower photographs of the entrance (white arrow), with a vertical shaft just big enough for a person.

Larger scale mining took place on the surface and below ground on the southern slopes of Needle Hill. These scars are the remnants of wolframite mining along quartz veins.

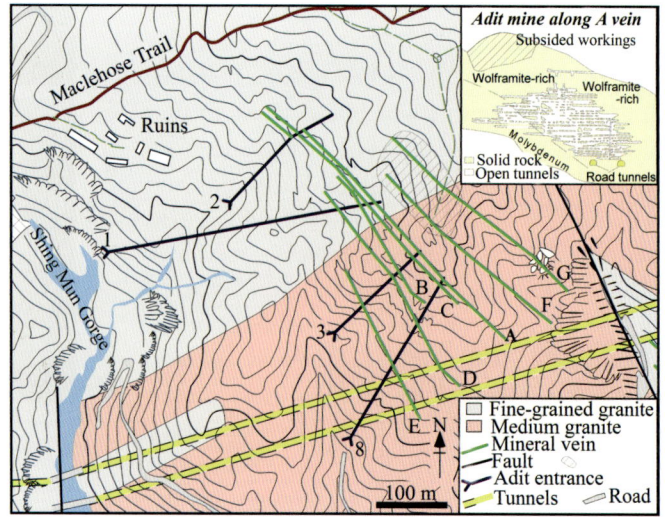

Wolframite was mined along several NW-trending quartz veins that cut through the Needle Hill Granite. These were labelled A–G. The extent of mining operations along Vein A is shown in the inset, which indicates just how close (within 5 m) the Shing Mun road tunnels are to the workings.

The Central New Territories: Human Impacts

Reclamations in the Sha Tin Valley

Human impacts accelerated with the arrival of the Kowloon-Canton Railway, opened in 1910. During the early 1970s, extensive reclamation for a New Town (IN30) were carried out, turning the former estuary of Tide Cove (see map, p. 116) into a major conurbation. However, there were beneficial ecological effects on the adjacent hills due to the cessation of grass cutting and charcoal burning as the economy changed. Environmental improvements were also furthered by the introduction of Country Parks (IN28, p. 112) and Special Areas, such as Tai Po Kau (IN31, p. 116).

Sha Tin dates back to the 1970s. All of the land occupied by the tall buildings in the photograph has been reclaimed. Note the railway in the foreground, which was constructed in 1910 and originally ran close to the former northwestern coastline. The small hill surrounded by buildings on the other side of the Shing Mun River is Yuen Chau Kok, which once lay within the waters of Tide Cove and was attached to the coast by tidal flats.

IN30 The New Town Programme

Increasing population pressures and a severe shortage of land around Victoria Harbour prompted planners to consider developing rural sites in the New Territories. New Town development in Hong Kong was first examined in a 1938 report by the Housing Commission, which was set up to investigate issues relating to slum clearance.

The Commission recommended the relocation of residents from decaying urban areas to several sites in the New Territories, including Sha Tin, Tsuen Wan, Yuen Long, Tai Po, and Fanling. The recommendations were not implemented, and, by 1971, the housing situation had reached a crisis, with 1.5 million people living on about 3,000 hectares in Hong Kong-Kowloon. Consequently, in 1973, the government embarked upon the New Town Programme, which planned to house up to 1.8 million people in six (later eight) New Towns by the mid-1980s.

The New Town Programme had a major impact upon the landscape of Hong Kong, involving extensive land formation and the building of roads and urban facilities, including parks and urban landscaping. Most of the residents are housed in residential blocks, up to 39 storeys high, to optimize land use and create an open living environment.

Today, the eight New Towns occupy 8,800 hectares, comprising about 3,000 hectares reclaimed from the sea. Development was rapid, so that by March 1995 almost 2.6 million people had been relocated to the New Towns, and, by the end of 2004, the total New Town population was about 3.2 million, almost double the original target.

IN31 Tai Po Kau

The Tai Po Kau Nature Reserve was designated in 1977, and consists of 460 hectares of plantations, some dating back to 1926. There are over 100 tree species and a wide variety of plants. Common trees include Chinese red pine, China fir, Brisbane box and camphor, which have matured and provide rich habitats. Over 160 birds have been recorded, including, for example, great barbets and grey-throated minivets. Birds of prey include crested goshawk and black bazza. Two-thirds of Hong Kong's butterfly species are found in the reserve, as well as many dragonflies, freshwater fish, and reptiles. Mammals include macaques, barking deer, pangolins, civets, and wild boar.

Sha Tin and Tuen Mun are the only New Towns built on virgin sites, both requiring extensive reclamation. Interestingly, reclamation in the Sha Tin Valley was examined as early as 1923, and in 1939, but the first Outline Development Plan was not drawn-up until 1954.

Development was carried out in stages, and the scope continually expanded. Plans for Stage 1, on the southeast side of the valley, were approved in 1965. The initial target population was for 145,000 residents by 1976. The final design was approved in 1977, by which time the target population had increased to 550,300 in Sha Tin and 80,000 in Ma On Shan.

From the early 1970s to 2004, Sha Tin and Ma On Shan grew from 29,000 residents to 628,300, with a planned population of 641,000 by 2005. In 2003, Sha Tin housed 9.2% of Hong Kong's population. Together, the towns occupy 2,000 hectares, with most land formation in Sha Tin being complete by 2004. Development at Ma On Shan is continuing, but is constrained by the presence of subsurface marble (IN32, p. 118).

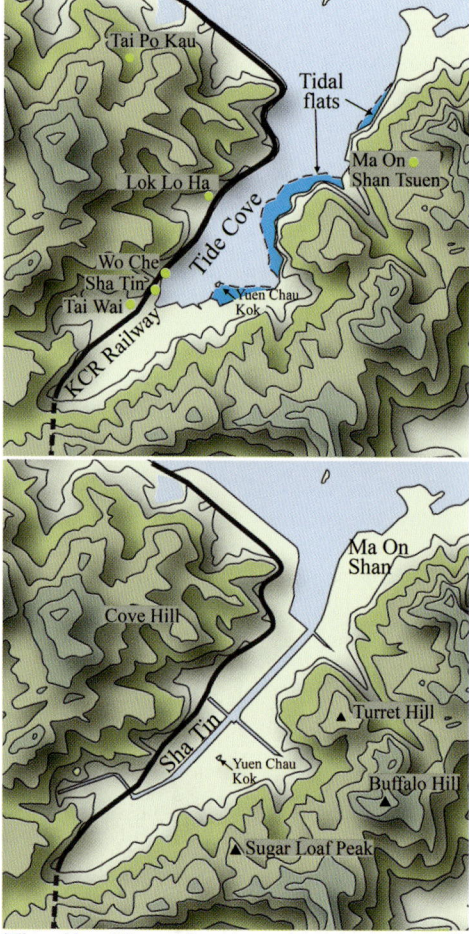

The Sha Tin valley has changed dramatically. The upper map shows the coastline as it was in 1957, and the lower shows the shoreline in 1999.

The Central New Territories: Human Impacts

Reclamation within the Sha Tin Valley has radically changed the landscape of the area. The waters of Tide Cove (above) have been filled in, and an extensive urban area has developed (below). The photograph above shows the former coast near Lok Lo Ha on the northern shoreline, with the twin peaks of Ma On Shan (right) and Ngau Ngak Shan rising in the background. Although the summit of Ma On Shan is composed of tuffs, most of the surrounding hills to the southwest, west, and north of the former Tide Cove are granitic. Quartz is abundant in these rocks and is probably responsible for the sandy beach in the old photograph. The photograph below shows the view from a similar, but slightly higher, location today that overlooks the new sewage treatment plant and New Town developments on the south side of the Shing Mun River.

IN32 Scheduled Areas

Urban developments are an increasingly important component of the modern landscape. The siting of these built-environments is not a random occurrence but reflects a host of natural controls that were considered by the developers. Some controls are clear, such as the historical location of the city next to Victoria Harbour or the preference for flat ground rather than steep slopes. However, other factors that influence the siting of towns and cities are not always so evident, or even visible. For example, the geological conditions below the ground surface play an important role in determining where tall buildings can be economically located. In Hong Kong, areas characterised by particular problems are subject to planning, design, and engineering restrictions.

There are three geologically determined Scheduled Areas in Hong Kong. Area 1, in the Mid-Levels, is underlain by colluvium (IN17, p. 91). Area 2, the northwestern New Territories, and Area 4, Ma On Shan, are underlain by marble (IN11, p. 51). Importantly for foundation design, marble has a highly irregular surface, solution cavities, and widened joints that weaken the rock and present engineering difficulties. Areas 3 and 5 are linear zones that are drawn 30 m either side of the alignments of the MTR and sewage disposal tunnels, respectively.

The Northshore Lantau Designated Area also has marble below the surface and, in addition, very deep weathering, with soft sediment infilling depressions in the weathering front (IN51, p. 177).

Difficult ground conditions in these areas require that specialists supervise the drilling. If marble is encountered in a borehole, the driller is required to penetrate at least 20 m into good quality rock to minimise risks from solution cavities. If voids are encountered, the regulations demand that drilling must continue to even greater depths.

Three areas are underlain by marble with cavities: Yuen Long, Ma On Shan, and Northshore Lantau. These have complex geological conditions that present problems for urban development.

THE SOUTHEASTERN NEW TERRITORIES

Nearshore islands, a large bay, and a complex coastline, backed by high mountains, give the southeastern New Territories a unique maritime setting. The region extends southwards from Ma On Shan Country Park to the Clear Water Bay Peninsula. To the east, it is bounded by the Sai Sha Rd. and, to the west, by the Fei Ngo Shan ridge. Ngau Mei Hoi (Port Shelter) is a large body of water that constitutes half of the total area and encompasses many islands, the three largest being Kai Sai Chau, Tiu Chung Chau, and Kiu Tsui Chau.

The most prominent features in the landscape are the mountains of Ngau Ngak Shan (the Hunchbacks; 674 m), Ma On Shan (Horse Saddle Mountain; 702 m), Shui Ngau Shan (Buffalo Hill; 606 m), and Fei Ngo Shan (Kowloon Peak; 603 m). All are composed of hard volcanic rocks that have generated a rugged landscape with several dramatic ridge lines.

The higher mountain slopes were once cultivated, but the terraces are now abandoned. Today, they are characterised

High ground links several major peaks in the northwest of the area, with a spine of land extending to the southwest along the Clear Water Bay Peninsula. The west-east profile is viewed from the southeast.

Ma On Shan Country Park. The Hunchbacks and Ma On Shan lie on the horizon with the Ngong Ping Plateau to the right and in the middle distance. The left peak is Shek Nga Shan as seen from Buffalo Hill.

Woodland Path, near Mui Tsz Lam, Ma On Shan Country Park

by fire-maintained grasslands (mainly duck-beak grass and minireed). Broad-leaved scrub occurs on the lower slopes and in sheltered areas. Tall scrub and woodland tend to be found on the northern, less-sunny slopes, where moisture remains longer in the shady conditions. Extensive, native, broad-leaved forests extend over the northern slopes of Ma On Shan, the Hunchbacks, and Buffalo Hill (p. 120). These densely wooded areas also support populations of wild boar and barking deer.

Geological Background

The rocks in this region (map below) are mostly igneous in origin, having either crystallised below ground in large magma chambers, or erupted explosively as ash from volcanoes. The first rocks to form were the Needle Hill Granite and Sha Tin Granite. The magma from which they originated cooled and solidified slowly (over many tens of thousands of years) within magma chambers that lay deep within the crust. These events took place about 146 million years ago (figure, p. 107), during the second phase of back-arc volcanic activity that affected Hong Kong (IN23, p. 105).

During the third intrusive phase (about 142 million years ago), the Shui Chuen O Granite was formed, again in a large magma chamber below the surface. At

Several cliffs composed of tuff occur on Fei Ngo Shan (photographs above). Large areas of the southeastern New Territories are dominated either by rocks like these (map below), or by granite that formed in magma chambers deep below the ground.

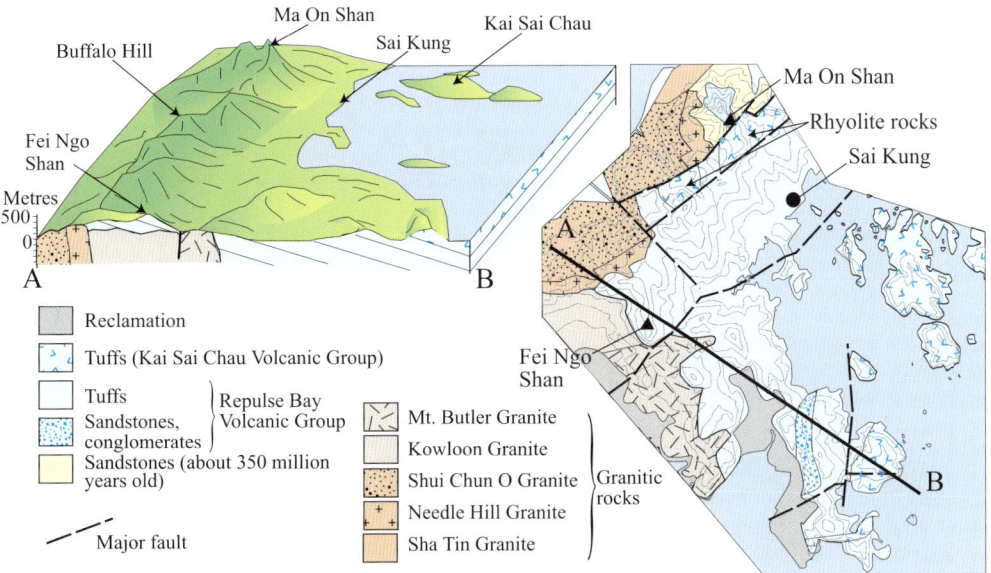

The Southeastern New Territories: Geological Background

about the same time, violent eruptions at the surface ejected tuffs of the Repulse Bay Volcanic Group. Some of these rocks accumulated within collapsed volcanoes, or calderas, that may or may not have been linked together (IN40, p. 136). Locally, the volcanic materials were eroded by streams and were re-deposited as conglomerates, sandstones, and reworked tuffs (tuffites).

The fourth, and final, phase of intrusion occurred about 141 million years ago, resulting in the formation of the Kowloon Granite and the Mt. Butler Granite. At the surface, eruptions of lavas and ash took place, from a fissure located in the area of the present day Sai Kung Country Park. This activity produced the material that constitutes the Kai Sai Chau Volcanic Group, which dominates the Clear Water Bay area. At about the same time, near Buffalo Hill, fissure eruptions (adjacent figure) occurred along ring faults (IN40, p. 136) and produced rhyolite dykes (IN04, p. 22). These dyke rocks can be traced into rhyolite lavas that flowed over the former land surface. The rhyolites are relatively hard and resist erosion. Consequently, they have formed several rugged ridge lines and rocky outcrops.

Delta sandstones, about 350 million years old, make up part of the Bluff Head Formation

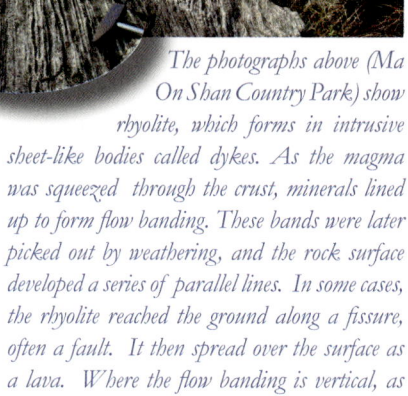

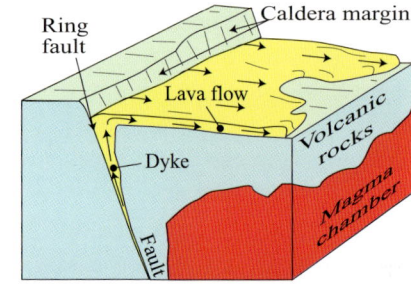

The photographs above (Ma On Shan Country Park) show rhyolite, which forms in intrusive sheet-like bodies called dykes. As the magma was squeezed through the crust, minerals lined up to form flow banding. These bands were later picked out by weathering, and the rock surface developed a series of parallel lines. In some cases, the rhyolite reached the ground along a fissure, often a fault. It then spread over the surface as a lava. Where the flow banding is vertical, as in the photograph, the magma probably cooled and solidified in a dyke Where the flow bands are nearly horizontal, it is more likely that the rhyolite formed as a lava flow (diagram below).

The photograph shows the landscape northwards from Shek Nga Shan near Buffalo Hill. The two rugged ridges in the foreground are composed of rhyolite that formed about 141 million years ago, which is the time when the rock formed and not the age of the ridges. These developed much later, probably in the last few hundred thousand to few million years and are the result of erosion acting on the hard resistant rhyolite. The prominent, rounded granite hill in the background is Nui Po Shan (Turret Hill) on the southern edge of Sha Tin.

The Southeastern New Territories: Geological Background

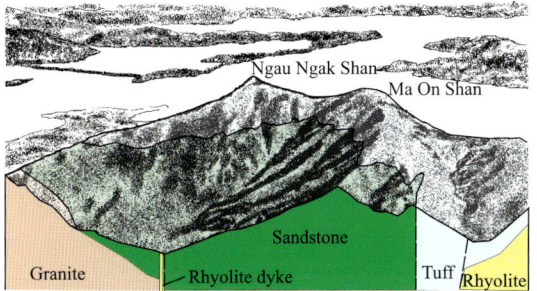

(IN05, p. 26). These rocks occur in the north, below Ma On Shan and Ngau Ngak Shan, the summits of which are formed by resistant tuffs (figure above).

The photograph above shows the challenging ridge walk along Ngau Ngak Shan (highest point), as viewed from Ma On Shan. The steep slopes are made of tuff. The lower, gentler slopes are mainly sandstones, as can be seen in the adjacent sketch.

Marble occurs below the western slopes of Ma On Shan, underneath the New Town and the Ma On Shan iron mine (IN33, p. 124). The iron deposits provided the base for significant mining activities in the past that have left many scars on the landscape (p. 129). The iron occurs in a lens-shaped, steeply sloping (35–55°), ore body dominated by the mineral magnetite (photograph right; figure below right).

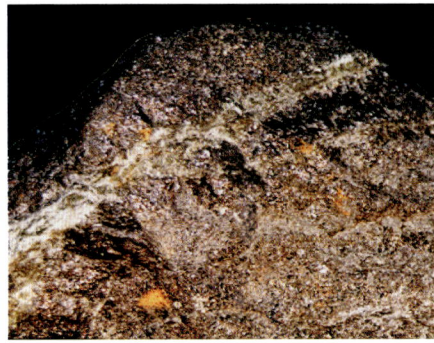

Magnetite (above) is by far the most common iron mineral in the Ma On Shan area and occurs as very small, black crystals scattered through the host rocks.

The magnetite (Fe_2O_3) was originally formed by a chemical reaction between granite magma and pre-existing marble. These reactions also produced a zone around the magnetite ore body with abundant and varied minerals. This is called a skarn. Remains of the original marble are exposed in the underground workings as a large, northward-dipping mass that occurs in the centre of the ore body (right). The adjacent Sha Tin Granite has been dated at 146 million years old. The skarn mineralisation would be of the same age.

Magnetite occurs within a steeply dipping central ore body surrounded by a mineral rich zone called a skarn that formed as a result of reactions between molten granite magma and calcium-rich marble.

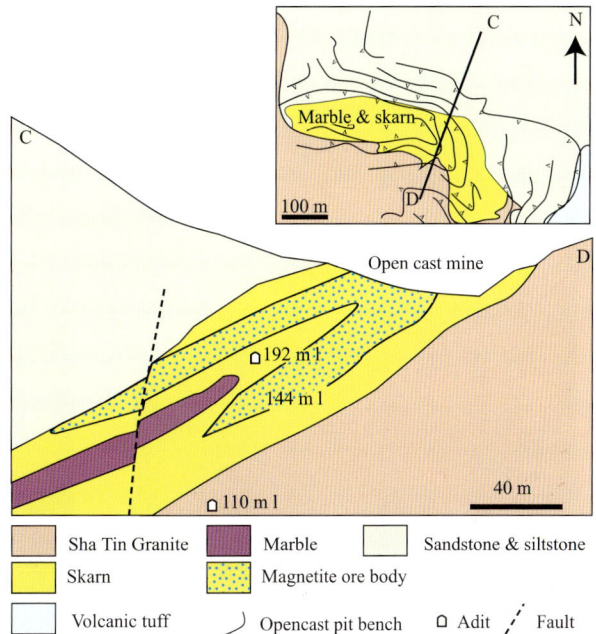

IN33 The Ma On Shan Iron Mine

The mine lies 5 km from Sha Tin. Ore was first reported in 1906, and a prospecting licence was obtained by the Hong Kong Iron Mining Company in 1906. In 1931, a 50-year Crown Lease was issued to the New Territories Iron Mining Company.

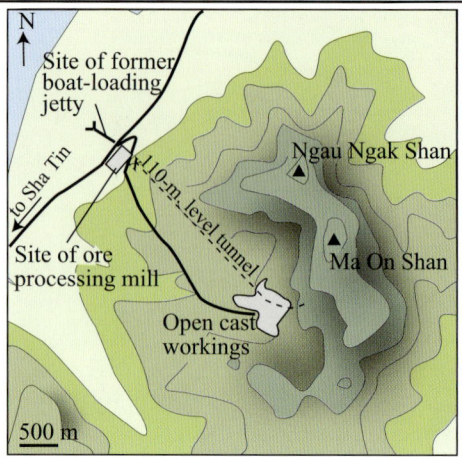

Annual production up to 1947 was low, at less than 1,000 tonnes. Output increased when opencast mining began in 1949, reaching 169,374 tonnes in 1950. Underground mining began in 1953 when the Mutual Trust Company signed an agreement with the Japan-based Nittetsu Mining Company Ltd. to provide assistance with operations. Opencast working ceased by 1959, and, by the early 1960s, all reserves in the upper levels had been exhausted. Development of deeper reserves began in 1963, comprising 5,458 m of main tunnels and shafts and 3,000 m of sub-levels. Costs were reduced by constructing a 2.2 km long, HK$3.5 million, haulage adit at the 110 m level. A new portal near the processing plant, only 200 m from the coast, avoided the long journey down the winding road from Ma On Shan Tsuen.

By the early 1970s, production increased to over 400,000 tonnes a year. Three adverse events occurred in the mid 1970s. A worldwide decline in the demand for steel, the opening up of large iron deposits in Australia, and the termination of a contract to supply Japan led to the suspension of mining in March 1976. The workforce of 400 were laid off, and the mine was abandoned in March 1981. Records show that, between 1949 and 1976, about 3 million tonnes of processed iron ore had been exported, mostly to Japan. About 4 million tonnes of ore remain.

Today, the opencast benches can be seen extending up to the 300 m level. Most of the dilapidated buildings survive, although they are overgrown. Portals at the 240 m level, near Ma On Shan Tsuen, and at the 110 m level, near the processing plant, can still be entered. However, although ventilation and drainage are generally still good, roof collapses have caused ponding of water in several sections. Cave-ins and rotten supports are common, and the vertical shafts present serious dangers. The tunnels should not be explored.

Sea Level Change and Islands

Over the last two million years, both climatic cooling (glacial periods) and warming (interglacial periods) trends have occurred, with ice advancing and retreating many times. From the peak of the last glacial advance, temperatures have increased by 5°C, causing ice on the land to melt and water to return to the oceans. Consequently, sea levels have started to rise. In some places, such as northern Europe, this rise was partly compensated by a simultaneous upward movement of the land as the weight of kilometres of ice was removed—a process called isostacy. In Hong Kong, there were no glaciers and the region felt the full effect of a rising sea.

These sea level changes had a profound affect locally, as can be seen in the maps to the right. About 18,000 years ago, sea levels were very low and a network of rivers radiated from the mountains of Hong Kong and flowed across broad open plains, before turning southwards towards the coast, which lay about 120 km to the south of Hong Kong. By about 8,000 years ago, the sea started to flood in along the former river valleys. This process continued for 2,000 years, by which time the coastline was similar to that of today. Although the sea may have briefly risen 2.5 m higher around 5,500 years ago, the sea level has since remained constant, and subsequent changes in the shoreline have been due to the infilling of flooded valleys by river sediments (IN34, p. 126) and man-made reclamations.

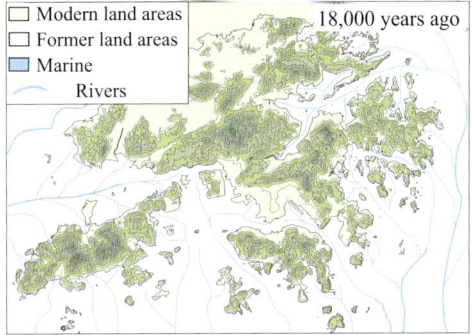
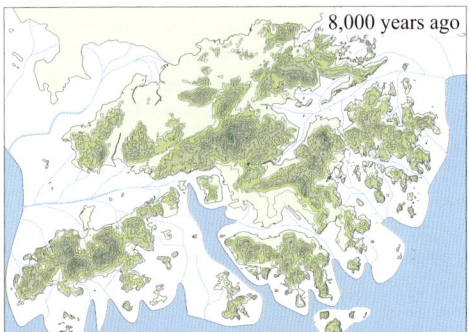
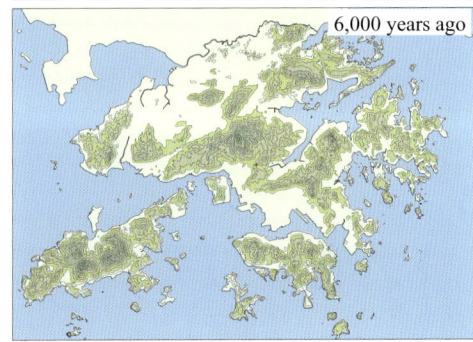

Sea levels have changed many times over the last few hundred thousand years in response to advancing and retreating glaciers on land. Sea levels were similar to today about 125,000, 210,000, and 350,000 years ago (figure below). These were interglacial periods. During full glacial stages (when ice reached its maximum extent), sea levels fell by about 120 m. The most recent glacial retreat started 18,000 years ago and resulted in the gradual flooding of Hong Kong to produce its highly indented coastline (maps above).

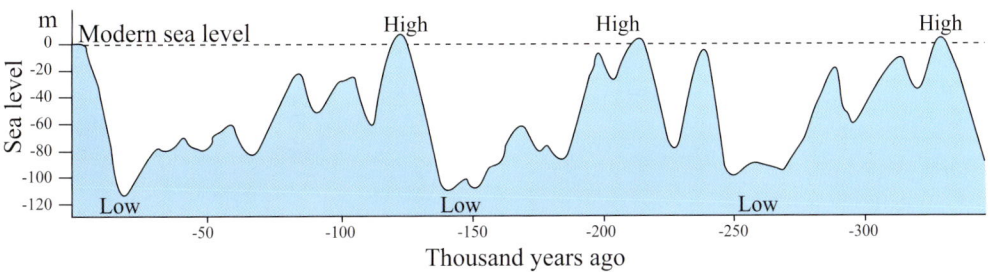

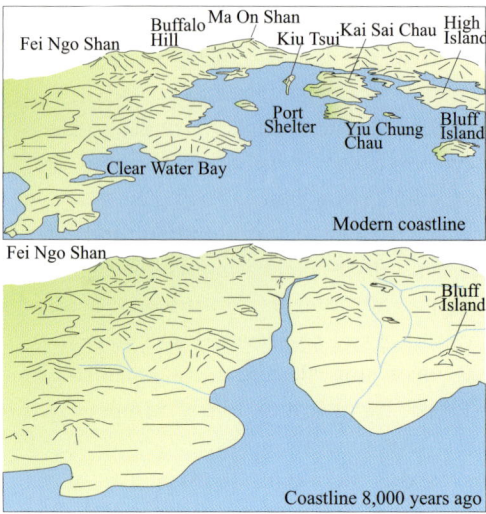

The southeastern New Territories were profoundly affected by these changes. Islands and the inter-island channels in the area simply represent the tops of mountains and river valleys that have been flooded. The sketch to the right shows how this offshore landscape has changed since about 8,000 years ago.

The sea level fluctuations that took place in Hong Kong are only part of a more widespread, world-wide phenomenon. Future changes in sea level remain uncertain, but given that the difference between the last ice advance and today is only 5°C, the potential dangers of human-induced global warming are clear.

Islands such as Kai Sai Chau and Kiu Tsui (photograph and upper figure) have only existed for about the last 6,000 years. Prior to 8,000 years ago, the area was entirely land, with the modern islands forming hills standing above a flat river plain. Rising sea levels began to flood the area from 8,000 years ago (lower figure).

IN34 Valley Deltas and Progradation

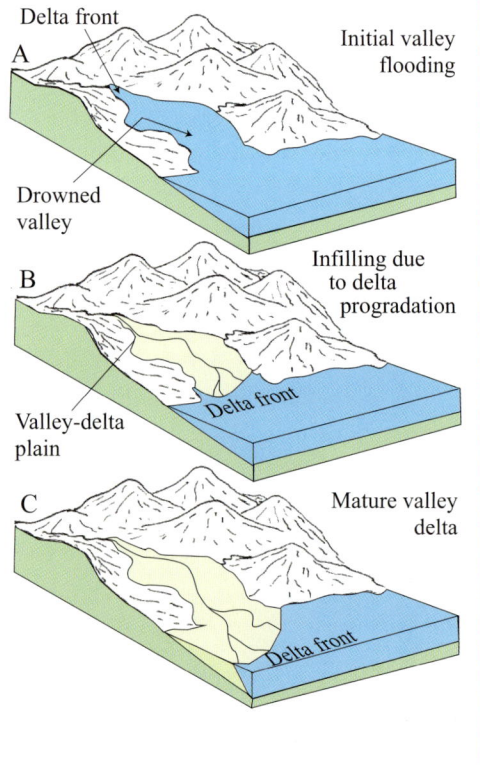

Sea levels began to rise rapidly with the melting of glaciers after 18,000 years ago, reaching a maximum height, 2 m higher than today, 5,500 years ago. In broad terms, they have stood at or near modern heights for the last 6,000 years. Rising sea levels resulted in the flooding of former river valleys (A). These drowned valleys were protected, in most cases, from wave action. This, plus the limited tidal range in Hong Kong, brought about calm, low energy, conditions favourable for sediment to settle to the sea floor. Large quantities of sediment in the rivers, due to heavy summer rains, allowed rapid infilling of the drowned valleys, and the coastline advanced seaward, leaving flat delta plains across which rivers flowed (B). This process is called progradation and, eventually, leads to a mature stage (C) when deltas begin to spread out from their valley bays.

Human Impacts

Upland terraces occur in several areas. These were built by former Hakka settlers, who had to occupy higher ground because the more productive lowlands were already settled by the Punti (Cantonese). The terraces, in areas such as Mau Ping and Wong Chuk Shan (right photograph), were once used to grow rice (where irrigation was available), tea and indigo (on the higher slopes), and vegetables and peanuts (on the lower slopes). Many are now overgrown with grasses or lost amid regenerated forests.

Stone masonry skills are evident in the well-laid stone paths and terrace walls. A network of paved trackways crosses the hill ranges here and in other parts of Hong Kong. Many of these routes were built in the reign of Qian Long (1736–96 AD), both to provide local access to markets, such as Sai Kung, and to integrate into a regional north-south network of trade routes that stretched to Canton.

During the nineteenth century, Hakka villagers were dressed in hemp cloth. Fibres from the stems of

Many upland villages, such as Wong Chuk Shan (above), have been abandoned over the last few decades and are now derelict, with trees reclaiming the landscape. The skill of the former villagers is often shown by the numerous terrace walls (below) with interlocking stones that are still standing many years after abandonment.

the locally-grown plant were stripped and spun into thread by indigenous women. Travelling groups of Hakka males then took the thread and wove it into cloth, which the local women dyed and turned into garments.

Although many villages have long been abandoned, local traditions still survive in the form of well-maintained graves that are dotted across the hillsides. Many are old, but new ones continue to be constructed for those with rights to traditional burial sites (IN35). These armchair-shaped structures are a distinctive cultural aspect of landscape throughout Hong Kong.

Traditional Chinese graves are present on many hillsides. Some occur in clusters, but others stand alone, often with spectacular views of the surroundings. Most are well-maintained by descendents who regularly sweep the graves and burn offerings.

IN35 Chinese Graves

Traditional Chinese beliefs are based on the existence of yin (female) and yang (male) characteristics. The moon is said to be yin, the sun to be yang. A living person resides in the yang world and upon death enters the yin world. It is believed that those in the yang world should attempt to maintain a harmonious relationship with the yin world. People in the living world reside in yang houses, whereas those who have died live in yin houses or graves. The need to maintain harmonious relationships leads people to build stylish yin houses (graves), to care for graves, and to make offerings of food, paper money, and other items.

When a person dies, the family will often consult a Feng Shui master to determine where, when, and how the body should be buried. Families come together at the graveside during the Ching Ming and Chung Yeung festivals to sweep the grave, worship the ancestors, burn paper offerings, and remove overgrowth.

Traditional practises in Hong Kong (and southern China) involve reburial, with remains being exhumed after several years. Upon removal, the bones may be cremated or cleaned and placed into urns (jinta). These are often located on a hillside or in a small concrete shelter with a favourable open view and good Feng Shui. These jinta shelters require official approval, which is often restricted. Today, most urban dwellers are cremated, and their ashes are placed in a niche in a public columbarium. Graves vary in shape, but most resemble an armchair, which symbolises authority and power, as armchairs were formerly the seat for magistrates when presiding in court.

The Southeastern New Territories: Human Impacts

Tropical rainforests once covered most of the area, but these were cleared by early settlers, who created a patchwork of new habitats, ranging from montane grasslands, through paddy fields, to scrublands, montane forests, lowland forests, and Feng Shui woods. These environments are now protected within the Ma On Shan Country Park (IN28, p. 112) and support a considerable biodiversity that includes, for example, a wide range of butterflies (IN37, p. 130).

During the twentieth century, mines and quarries have been added to the landscape. Slope instability has developed at some of these sites, as a result of undercutting of slopes. Large landslip scars (top right), for example, occur on Ma On Shan, above an old iron mine (IN33, p. 124) that weakened the adjacent hillside (figure). At other places, quarries have been used as landfills (IN36), all of which have been, or are scheduled to be, landscaped.

Landslip scars (white arrows) are common on the mid-slopes of Ma On Shan. These formed following opencast quarrying for magnetite (figure below) that undercut the slopes. Support was lost and slumps (IN45, p. 165) developed. These are blocks of material that have rotated to produce curved headwall scars and toe regions of disturbed material (figure below).

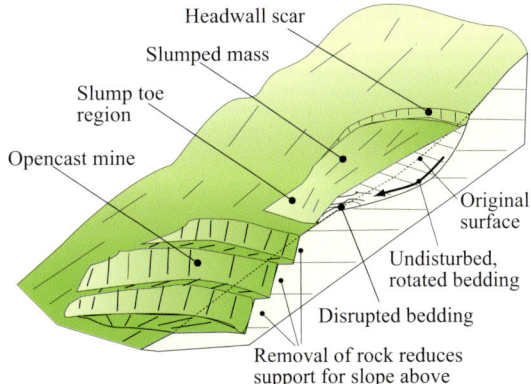

IN36 Landfills

Landfills are unsightly, and appear unnatural, even after restoration. Three were built in the 1990s at a cost of HK$10 billion and designed to last until 2020. By 1999, Hong Kong was dumping some 18,000 tonnes per day, exceeding the planned rates of infill and reducing the landfills' collective working life to about 2015. The sites are located at Tai Chik Sha, Ta Kwu Ling, and Nim Wan. The Tai Chik Sha site (100 hectares; 43 million tonnes capacity) lies on the western Clear Water Bay Peninsula and will be filled to a depth of 100 m with waste. The Ta Kwu Ling site (61 hectares; 38 million tonnes) lies in the northeastern New Territories and has a maximum infill of 140 m. The site is Nim Wan (110 hectares; 61 million tonnes) lies on the shores of Deep Bay and has a maximum planned waste depth of 120 m.

There are 13 closed landfills that are in various stages of restoration. The photo shows the Phase II/III landfills at Tseung Kwan O (near Tai Chik Sha). Most are due to serve recreational functions.

IN37 Butterflies

Butterflies are one of the major attractions of the Hong Kong countryside, with hundreds being visible at times. They demonstrate the high biodiversity that typifies much of the territory, with about 240 species from ten distinct families. The total number of species is equivalent to some 15% of all the butterflies present throughout the whole of China.

In general, they are most abundant between April and June, and from October to November. Although not as common in winter, certain species can sometimes be seen after the sun has warmed the temperatures to above 22°C, below which they tend to be inactive.

Butterflies are winged flying insects. Like other insects, they have a head, thorax, and abdomen, plus six legs, a pair of antennae, and compound eyes. The body is covered by miniscule sensory hairs.

Both butterflies and moths form part of the order Lepidoptera. Lepidos is the Greek word for "scales", and ptera means "wing". Butterflies differ from moths in several ways. The former are active during the day and are characterised by club-like antennae and bright colours. They usually rest with erect wings that are folded together. In contrast, moths are nocturnal and have feathery antennae, dull colours, and rest with wings that are open or folded over the back.

Red Ring Skirt
Hestina assimilis

Rose Helen
Papilio helenus

Large Faun
Faunis eumeus

Striped Blue Crow
Euploea mulciber

Common Sergeant
Athyma perius

Banded Tree Brown
Lethe confusa

Common Five-ring
Ypthima baldus

Great Orange Tip
Hebomoia glaucippe

THE EASTERN NEW TERRITORIES

The eastern New Territories possess some of the most varied, and in places the most exciting, coastlines in Hong Kong. The region is almost entirely made up of volcanic rocks, some of which were formed by mega-eruptions associated with the collapse of long extinct volcanoes. The region stands in splendid isolation from the rest of Hong Kong, being linked only by a narrow strip of land. The area consists of a large peninsula that juts out towards Mirs Bay bounded to the north by the Tolo Channel and to the south by Port Shelter. There are no towns, although several small villages are scattered across the region. The latter were intentionally excluded from the Sai Kung East and West Country Parks (IN28, p. 112), which protect the rest of the area.

Here the coastline is a highly indented and flooded, or ria, landscape in which river valleys have been inundated by rising sea levels. The relatively calm waters of Long Harbour (Tai Tan Hoi Hap) cut deeply into the region following one such ancient river bed. Several flooded valleys (Yung Shue O; Sham Chung) have partially been filled with sediment brought down from the eroding mountains. High Island Reservoir lies in the south,

The eastern New Territories is a land of grassy mountains and sea views. Sharp Peak (below, centre) is the most distinctive, and isolated, summit and makes a challenging hike up its steep and eroded footpaths.

The Eastern New Territories

with its three dams that now link the formerly isolated High Island to the rest of the eastern New Territories.

The summits are mostly broad and rounded, and all are less then 500 m, lower than many other parts of Hong Kong. The highest mountains include Shek Uk Shan (484 m), Sharp Peak (468 m), Tai Cham Koi (408 m), and Kai Kung Shan (399 m).

The vegetation cover changes gradually across the region, with woodlands in the west giving way to grassy slopes eastwards, reflecting the increasingly drier conditions (IN38). The forests are mostly plantations, some being grown following government efforts, with others resulting from the work of villagers in the 1950s. Native broad-leaved trees are slowly

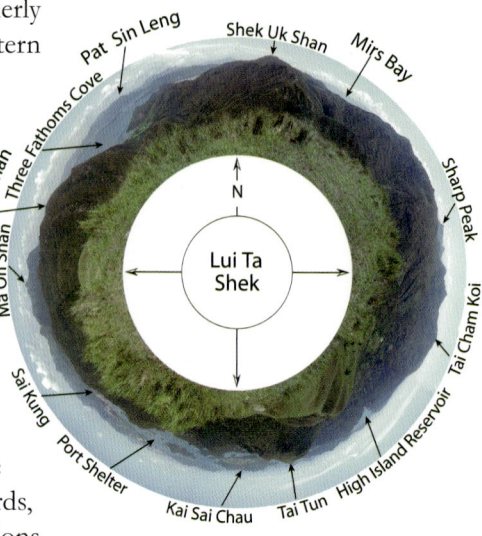

Many of the hills in this area offer superb panoramas that extend over the mountains and surrounding waters. This view was taken from Lui Ta Shek in the Sai Kung West Country Park.

IN38 Hong Kong Weather

Hong Kong experiences a humid, subtropical, monsoonal climate with a pronounced seasonal rainfall. The winter is cool, dry and sunny. Average daily temperatures range from 13°C at night (January and February) to 25°C during the day in November, and 19°C in January and February. The lowest winter temperature reading at the Hong Kong Observatory was 0°C on 18 January 1893. Colder temperatures occur on high ground, such as the summit of Tai Mo Shan, which occasionally experiences frosts. Monthly rainfall ranges between about 26 mm (December) and 45 mm (January). This is the time of year when the danger of hill fires is at its greatest.

Summer is humid and bright, with an average of 7 hours of sunshine a day. July is the hottest month, with daily shade temperatures of 26–32°C. The highest recorded summer shade temperature at the Hong Kong Observatory was 36.1°C

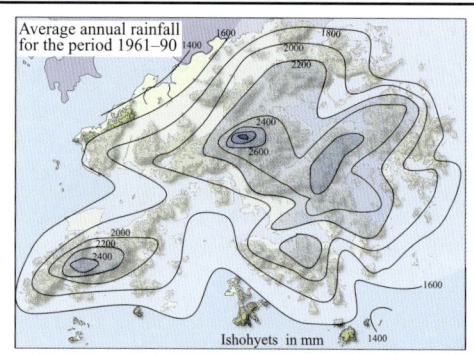

on 19 August 1900. Almost 80% of the annual average rainfall of 2,124 mm (at the Hong Kong Observatory) falls in summer. June and July are the wettest months, with June receiving an average of 476 mm.

Typhoons can occur at any time between May and November, but are most common from July to October. Typhoons bring gale force winds and heavy rain. Flash floods are a feature of the steep mountain streams, which can quickly fill up to dangerous levels with the onset of intense tropical rainstorms.

replacing the main plantation tree, Chinese red pine, which has had serious problems with worm infestations.

The largest animals are feral cattle. Most of the wild animals are nocturnal and consequently are seldom seen. Wild boar, Chinese leopard cats, porcupines, pangolins, masked palm civets, and pythons have all been reported. Birds vary in abundance with the habitat, being less frequent on the grasslands. Butterflies, moths, termites, dragonflies, and other insects (IN39, p. 134) are common, reflecting the rich variety of species that occur in many parts of Hong Kong.

The largest animals roaming the countryside are feral cattle, remnants of herds once owned by the local villagers and now abandoned.

Rocks and Scenery

The rocks in the area are almost entirely of volcanic origin. However, they were formed at slightly different times. The oldest are tuffs of the Shing Mun Formation (164 million years old; p. 106) in the extreme northwest. These are overlain by sediments belonging to the Lai Chi Chong Formation, which were eroded from the older volcanoes, transported in rivers, and then laid down in a shallow lake about 146 million years ago.

A second series of volcanic sediments (the Mang Kung Uk Formation) occurs in the southwest. These were also deposited in a shallow lake setting somewhat later, about 142 million years ago. Subsequently, significant volcanism took place as part of the third major phase of igneous activity that affected Hong Kong (IN23, p. 105). This resulted in volcanic ash accumulating within a large

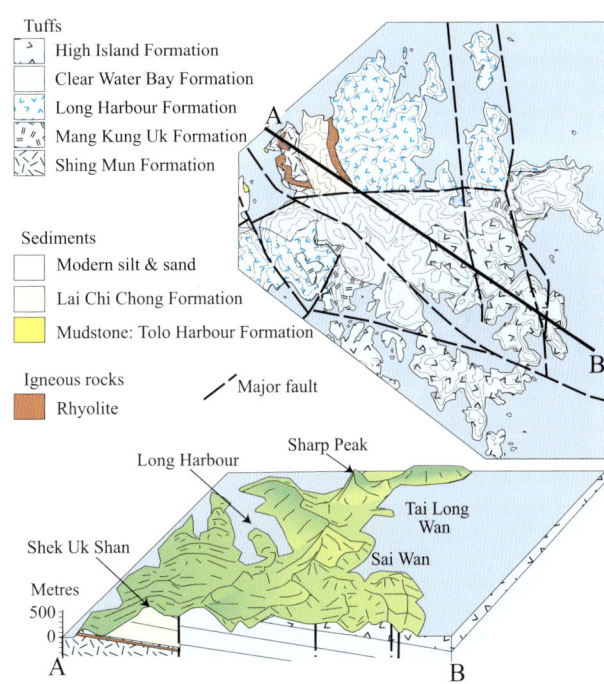

The region is dominated by volcanic rocks formed either as lavas that flowed over the surface or as ash that was blown violently out of volcanoes. The oldest rocks occur in the northwest, and the youngest are present in the southeast, reflecting a general trend for igneous activity to migrate southeastwards. The area is also crisscrossed by several faults that control the orientation of the modern valleys and bays, For example, note how Long Harbour is aligned parallel to a N-S-trending fault.

IN39 Insects

Over one million insect species have been recorded worldwide, and some specialists suggest that there may be more than 10 million. In Hong Kong, there are about 230 species of butterflies alone (IN37, p. 130). Moths (A) are common, the largest being the atlas with a wing span of up to 30 cm. Caterpillars (B–D) are equally abundant. Dragonflies (E, F) are notable along many streams, with more than 110 species occurring, some of which are unique to the region. Spiders are frequently seen, especially the large woodland spider (G) with webs up to 3 m in diameter that stretch across many trails. Grasshoppers (H) are often present on the grasslands.

Many insects have an impact on landscape. Termites, for example, play an important role in soil processes. The most common, local species is *Coptotermes formosanus*. *Capritermes fuscotibialis* is also abundant, along with *Macrotermes barneyi*. Because of the cold winters, local termite species build their tunnels underground with low, laterally

extensive mounds and tubes present at the surface (photograph left). These are generally only visible after a hill fire.

caldera (IN40, p. 136) that lay to the north of an east-west aligned fault. These deposits now comprise the Long Harbour Formation. The final phase of igneous activity (IN23, p. 105) took place about 141 million years ago and gave rise to the rocks of the Clear Water Bay Formation and the High Island Formation. The former consist of rhyolite lavas (IN02, p. 15) that flowed from a long fissure. These were relatively sticky and travelled only short distances from the eruption sites, probably no more than 10 km.

The High Island Formation is present in the southeast, and is composed of tuffs, originally erupted as ash. These materials accumulated within a large volcanic depression (the Sai Kung Caldera), cooled slowly, and contracted to produce columnar joints (IN40, p. 136). Today, these rocks control the intriguing shape of the spectacular cliffs along the southeast coast. Volcanic activity finally terminated at about this time due to the cessation of the driving force, subduction (IN23, p. 105).

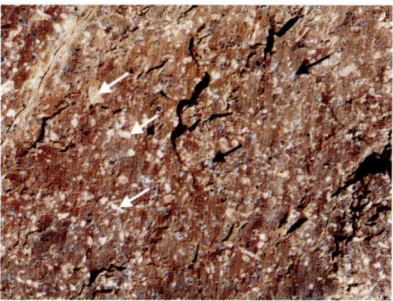

The rocks of the Clear Water Bay Formation (upper) are dominated by dark lavas with large crystals (called phenocrysts) of grey feldspar (white arrows). In contrast, the tuffs of the High Island Formation (lower) contain large crystals of both feldspar and quartz (black arrows).

The southeast coast of Sai Kung East Country Park consists of rugged cliffs cut into columnar jointed tuffs of the High Island Formation. These columns originally would have been nearly vertical, but have been tilted slightly by earth movements.

IN40 Calderas and Columnar Joints

Most volcanoes possess a broadly circular summit depression called a crater. However, in some cases, much larger circular features called calderas may develop. Several of these large structures (top right diagram) formed in Hong Kong during repeated phases of past igneous activity (IN23, p.105). Calderas are circular volcanic depressions, bounded by ring faults, along which magma may ascend, and along which collapse occurs. This commonly happens after large volumes of magma have been thrown out during an eruption. The overlying rocks then drop into the vacant space. Calderas may also form more slowly by gradual subsidence above a magma chamber, or by a cataclysmic explosion that blows the volcano apart.

The Sai Kung Caldera developed about 141 million years ago and extended across much of southeastern Hong Kong. It probably started through subsidence as lavas and ash of the Clear Water Bay Formation were erupted at the surface from a fissure vent (A). Subsidence probably continued over time producing several more faults that fractured the earlier rocks (B) and deepened the caldera. Then a major series of fissure vent eruptions (B) threw out very large volumes of hot ash, perhaps as much as 200 km^3 in a gargantuan eruption. This material ponded within the depression and cooled slowly, producing columnar joints (C) that today form such a distinctive element of the landscape in southeastern Hong Kong.

Columnar joints grow as a result of cooling of hot volcanic ash (or lava). Temperatures start to fall first at the surface of the hot ash layers. As the material gradually cools, it shrinks and hexagonal cracks develop that slowly extend downwards (adjacent figure). A similar process may also occur at the base of the ash layer, where it is in contact with cooler rocks below.

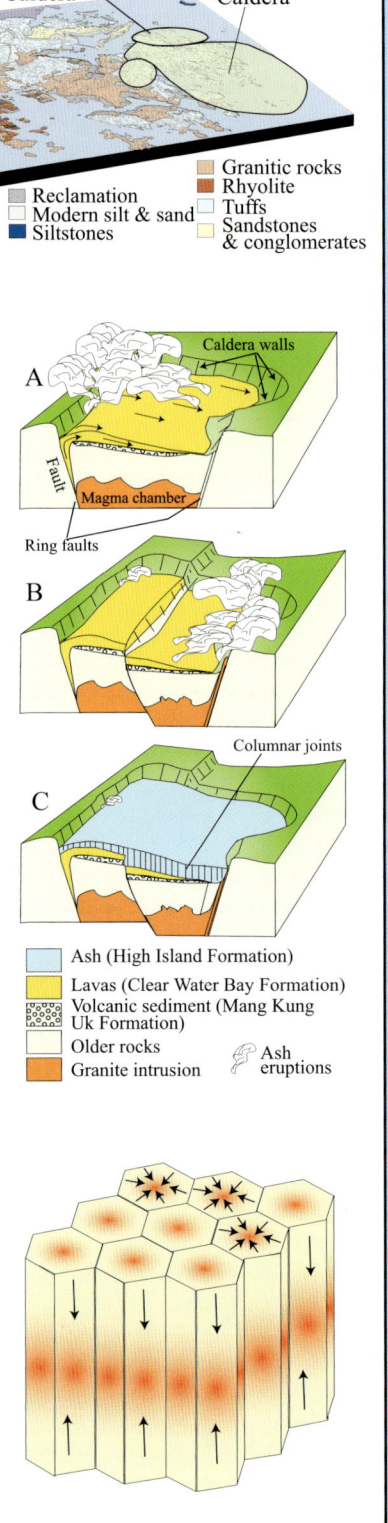

The Eastern New Territories: Coastal Environments

A number of small-scale intrusions occurred several millions of years after major volcanism ceased, in the form of basaltic andesite dykes (IN45, p. 155), which cut through the older rocks between 108 and 75 million years ago. These black rocks are usually less than 1 m thick and sheet-like in form. An excellent example (above) lies just below the High Island East Dam.

Columnar joints are sometimes bent as in this photograph. This occurred as a result of slippage (creep) downslope as tremors shook the caldera, and while the tuffs were still in a hot and semi-solid state. Note also the black dyke (arrows) cutting through the older rocks at a right angle to the columns. This is a basaltic (technically a shoshonitic lamprophyre) dyke that was intruded several tens of millions of years after the columnar joints were formed.

Coastal Environments

The coastline of Sai Kung is perhaps the most varied in Hong Kong. In the south and east, where the coast is exposed to heavy seas, there are cliffs and rocky coastlines. Where there is more shelter, e.g. at the head of bays, there are several beautiful beaches. Mangroves and muddy flats have developed in particularly calm locations, well-protected from the action of waves. Deltas have extended outwards from the land where rivers enter the sea.

Po Pin Chau is a sea stack visible from the High Island East Dam. Dramatic sea cliffs can just be seen from the dam, but they are best viewed opposite the side of the island that faces the South China Sea. The opposing shore consists of a steep slope covered with scrub, but no sea cliffs.

A long volcanic ridge lies to the north of Tai Long Wan (right in above photograph). Tai Long Tsui lies at the far end of the peninsula and is the easternmost point of the Hong Kong mainland. The area is exposed to powerful ocean waves, and several high sea cliffs have developed, together with a number of rugged clefts (joint slots), that cut deeply into the rocks along joint planes (photograph below). In some cases, erosion cuts through a headland to form a sea arch. Eventually, the roof of the arch collapses and leaves an isolated block of rock called a stack (figure below).

Erosional Settings

As waves approach a shoreline they "feel" friction with the sea floor and move more slowly. Furthermore, water depths are not constant, especially where there is an irregular coastline. Waves first encounter shallow water seaward of a headland and slow down (adjacent figure). The same wave in the adjacent bays will continue beyond the headland in deeper water, not slowing down because the sea floor is too deep to exert a drag. As a result, the wave will gradually bend towards the sides of the headland, concentrating its energy there and producing cliffs, caves, arches, and stacks. This process of wave bending is called wave refraction.

In addition, erosion will be greatest where softer rocks are exposed or where there are faults or joints (IN08, p. 35; IN09, p. 37). The Sai Kung shoreline is dominated by rocks composed of tuff and lavas. Initially, wave impacts erode the cliffs to form sea caves along weak

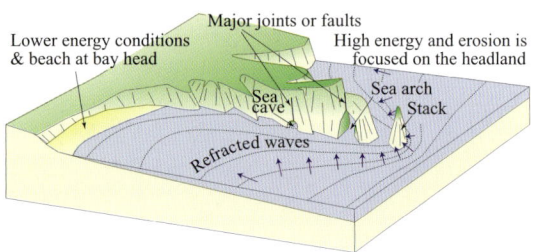

The Eastern New Territories: Coastal Environments

The classic sequence of landform development on a headland starts with erosion along joints or faults to form a sea cave (left). Then the cave is enlarged, and it cuts through to the other side of the peninsula to form a sea arch (centre). Eventually, the roof collapses and a sea stack develops (right).

rock zones, faults, or joints. Further wave erosion will enlarge these caves, which may cut through the headland or join up with another cave forming on the opposite side, and a sea arch begins to develop. Continued wave erosion will enlarge the arch until the roof can no longer support its own weight and it collapses, leaving an isolated column, or stack. Over time, this will be reduced to a low island and eventually disappear below the sea surface.

Beaches

Some beach deposits are derived from the erosion of cliffs and headlands. However, in most cases, the bulk of the sediment originates from the land and is brought down to the coast by rivers. Several distinct environments can be distinguished on most beaches, including aeolian (wind) dunes, backshore and foreshore zones (the "beach"), a shoreface, and the offshore zone just beyond the strandline region (IN41).

IN41 Beach Terminology

Beaches can be divided into several zones. Wind blown dunes occur above the average high-water level, where they are often stabilised by grasses. The backshore is only flooded during storms and often has wind blown ripples on its surface. Storms can throw sand above the high-water mark to form a flat, landward-sloping mound of sand called a berm. The foreshore extends between the high- and low-water marks and is dominated by swash and backwash (IN42, p. 141) processes that produce a smooth, flat surface. Locally, risers and treads (alternating steep and gentle slopes parallel to the shore) may develop as a result of breakers. The shoreface is the sub-tidal section extending from the low-water level to the fairweather wave base, which is the depth where waves can no longer move sediment. This depth is equal to half the wavelength of the waves that affect the coast. Beyond this point lies the deeper offshore environment.

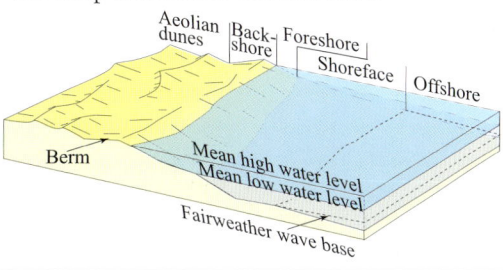

The Eastern New Territories: Coastal Environments

This small river at Lam Uk (northeastern Long Harbour) has been blocked and deflected northwards by a small beach (left). Blockage and drainage diversion is common and often brought about by longshore drift (IN42, p. 141).

As waves move sand particles, they determine the shape of a beach and the angle of its slopes. Gentle beaches develop on coasts where the waves are of a short wavelength (IN42, p. 141). As these wavelengths increase, steeper beaches, such as at Tai Long Wan, are formed. The distance between wave crests can also be linked to the type of breakers, with long wavelengths being associated with collapsing or surging breakers (IN42).

Many rivers approaching the eastern Sai Kung coastlines are deflected northwards by sand barriers. These are caused by a general tendency for mineral and rock grains along this coast to be pushed northwards as a result of longshore drift (IN42).

The eastern Sai Kung coast contains several spectacular beaches, with many diverting or blocking streams. This is the case at the three beaches shown to the right. Note the river diverted to the northern end of Long Ke beach (top), after bypassing the agricultural land of the Wui Oi Christian Training Centre. River blockage can be seen clearly in the photograph of Sai Wan's northern beach (middle). The stream partially reaches the sea by seepage through the sand and has formed a small freshwater lagoon. Three beaches lie at the head of Tai Long Wan (bottom). At the Ham Tin beach (foreground), stream deflection to the north can again be seen.

IN42 Beach Processes

The distance between two wave crests is called the wavelength, which has an important effect on the shape of beaches. Longer wavelengths are associated with more powerful ocean swells and taller waves. These drive sand up beaches, making them steeper. In contrast, short wavelength waves create gentler beach slopes.

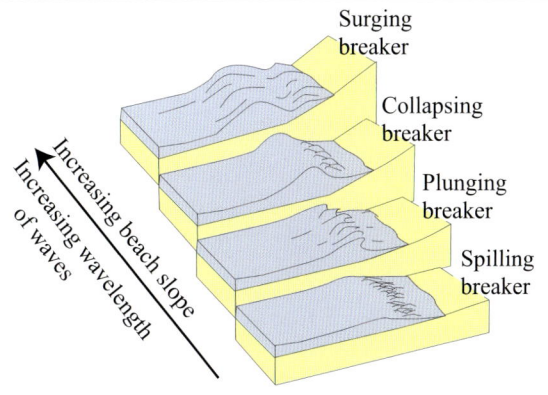

The kind of breaker in a surf zone is dependent on the beach profile. Spilling breakers develop where there are gentle slopes that allow waves to lose energy, gradually through friction. Other types lose energy more quickly as they hit progressively steeper beaches. Plunging breakers occur on moderate slopes and have a curling crest that drops over a pocket of air. Collapsing waves break in the middle of the wave front. Surging breakers form on very steep beaches. They do not break, but rather the base of the wave suges up the beach.

On steep beaches, rip currents may develop. These flow straight out to sea in a narrow zone perpendicular to the shore. The current moves at a high speed (up to 2 m a second), posing a danger to swimmers who can be dragged seawards. Rips occur when large approaching waves are parallel to the shore, and there is significant backwash from the collapse of previous waves. Rip currents are part of a larger circulation where water is moved onshore by waves and then flows parallel to the coast before returning offshore as a rip current. Rip currents carry sediment out to sea (photograph), and interfere with incoming waves, causing them to break early or not at all, producing a gap in the surf.

Beach sediment is moved on and offshore by swash and backwash. Swash is the onshore movement of water, whereas backwash is the return flow. When waves approach parallel to a coast, water simply moves up and down the slope. More commonly, the swash waves are oblique to the shore. The backwash returns directly down the steepest slope under the influence of gravity. Any particles carried by the water will first move up the slope at an angle with the swash waves and then return directly down the gradient with the backwash. This produces a net lateral movement of the sand, a process called longshore drift.

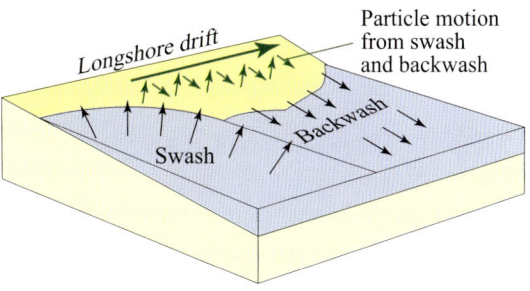

Long Harbour is shown above, looking northwards. This elongate bay was once a river valley, but it now represents an excellent example of a low-energy rocky coastline bordered by gentle slopes. The area is well protected from the influence of ocean waves. Consequently, no sea cliffs occur. The southern bays have shorelines characterised by mangrove stands, deltas, and narrow muddy flats. Tap Mun Island can be seen in the background.

Mangroves and Muddy Flats

Mangroves grow upon narrow, muddy flats in calm, quiet areas, with weak wave action, along the Sai Kung coastline. These coastal-marine trees and shrubs have shallow roots and prop roots that help to support the plants in the soft, unstable sandy mud in which they grow. Hong Kong lies close to the northern tropical limit of mangrove growth. As a result, small shrubby plants dominate, and plant diversity is low. There are eight common true mangrove species present in Hong Kong. These are dominated by *Kandelia candel* and *Aegiceras corniculatum*.

Dense mangrove stands occur on the Ham Tin River (below), and south of Ko Tong Hau in Long Harbour (above). Also, notice the birdsfoot delta developing in the extremely low energy setting of the latter example.

Deltas

Deltas (IN43, p.143) occur at several points along the coast of the eastern New Territories. For example, a series of small deltas occur at the southern end of Long Harbour. Wave action and sediment delivery by the inflowing streams are the main controls on the shape of the deltas. Where there is abundant sediment and little wave action, a birdsfoot delta tends to develop (upper right photograph). This occurs when sediments are dumped at a river mouth but are not redistributed by wave action. Consequently, the river extends itself seawards as a finger-like deposit. Eventually,

IN43 Delta Sediments

Deltas are lobes of sediment built up where rivers enter the sea. Rapidly flowing river water slows at the coast and dumps the particles it is carrying. Over time, accumulation of the deposits reduces the gradient of the river bed. The river then tends to move to a new position, a process called delta switching.

Several different areas can be distinguished on a delta (top right). Debris flows (IN46, p. 159), coarse-grained river sands, and boulders occur close to the source. Further out, the delta plain consists of pebbles and sands. Many deltas in Hong Kong also have a clay layer at the surface (right photograph) as a result of paddy farming.

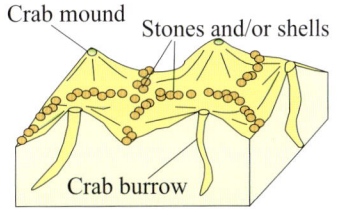

The intertidal zone is very varied, with the accumulation of shells, sands, and muddy sands. Wave action may drive sand bars (left photograph below) onto a delta. Living creatures also influence the nature of the sediments. For example, crabs often build small sandy mounds (right figure and photograph), with coarser gravels and shells accumulating between the mounds.

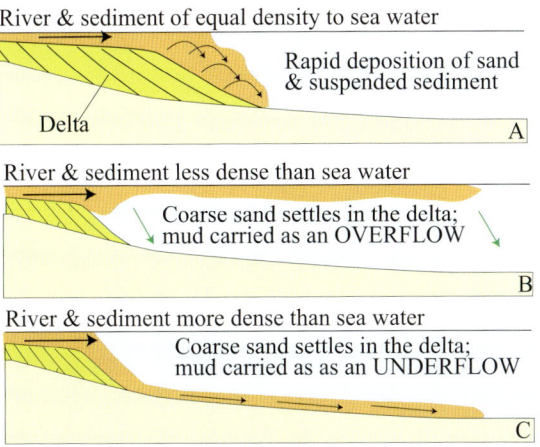

The delta front and prodelta are hidden below the sea. In these areas, sedimentation is controlled by fluid density. If the densities of the river and sea water are similar, then rapid mixing occurs and deposition takes place close to the delta, extending it seawards. Where the muddy river water is less dense, it will flow across the sea surface, carrying particles away. If the muddy river water is denser, an underflow forms.

A delta mouth near Chek Keng in Long Harbour shows the influence of waves, which have redistributed the pebbles (probably during stormy conditions) into a linear feature called a spit.

the gradient of the finger-like channel becomes very gentle, and the river finds a new route to the sea, forming a new finger. In contrast, fan-shaped deltas evolve when sediment is moved by waves. This spreads the river sediments along the coast. The direction of the prevailing winds and waves will determine whether the delta has a symmetrical or asymmetrical shape.

Coral Coasts

Coral communities (IN44, p. 145) are found mostly off rocky coastlines, both within sheltered and exposed bays and off remote islands in several parts of the eastern New Territories. They occur down to depths of about 10 m, although poor light conditions usually restrict vigorous coral growth to very shallow waters.

The Hoi Ha Wan Marine Park, with an area of 260 hectares, preserves several important coral communities. This park lies on the northern coast of the Eastern Sai Kung Country Park and was designated on 5 July 1996.

Corals are surprisingly abundant. In a 1982 study of shallow water (0.8–1.3 m

Scleractinian, or hard corals, such as the examples in the above photograph, are common in Hoi Ha Wan Marine Park. They occur scattered around the rocky coastline but are more abundant at four locations.

Hoi Ha Wan viewed from Mt. Hallowes (Tam Chai Shan).

deep) they were found to cover as much as 70–100% of the sea floor in parts of Hoi Ha Wan. However, the numbers of species and spatial extent appear to vary with time. A subsequent report in 1998 found a significant decline. The average coral cover was reported as 44% at 1 m depth but only 14% at 3 m depth. The extensive loss of corals in deeper water was attributed to a layer of cold water, with low dissolved oxygen and high salinity that developed in Mirs Bay in 1994. This may have resulted in up to 83% mortality. The natural event was caused by deep ocean water moving into Mirs Bay from the Luzon Straits. More recent studies have observed signs of a welcome recovery of the corals.

Human Impacts

Settlement of this area probably occurred later than in other parts of Hong Kong. The farmer immigrants tended to be of Hakka origin, who had to use the more marginal and less attractive locations for their homes. In 1960, the villages on the Sai Kung Peninsula consisted of 52 inhabited exclusively by Hakka, ten

IN44 Hong Kong Corals

Hong Kong corals form scattered communities rather than extensive reefs, partly because the local conditions are not ideal for coral growth. Temperatures are below those considered optimal. They are also more extreme, with especially cold conditions in winter and temperatures that can be too high in shallow water during the summer. Heavy rains and river discharges add fresh water to the saline marine ecosystem. Large inputs of silt and poor light penetration forces corals to live in shallow waters. Consequently, corals grow slowly and remain in balance with the processes that break them down (wave erosion; dissolution of calcium carbonate; burrowing organisms). In recent decades, pollution, dredging, reclamation, and dumping of mud have further inhibited coral growth.

There are two main coral communities in Hong Kong: Scleractinian and Octocorallion. The former are corals with hard skeletons that grow in northern Mirs Bay, Tolo Channel, and Port Shelter. In better conditions, they would form more extensive reefs. About 64 hard corals have been reported, including species belonging to the genera *Cyphastrea*, *Echinophyllia*, *Favia*, *Favites*, *Goniopora*, *Leptastrea*, *Lithophyllon*, *Pavona*, and *Platygyra*.

The Octocorallion (soft corals that do not secrete a hard skeleton) communities occur in the southeast, in areas such as Cape D'Aguilar, the Ninepins, Po Toi, and the outer islands of Port Shelter, such as Basalt and Bluff Islands. The latter locations are subject to powerful waves and strong tidal currents. Light penetrates more deeply in the clearer waters. Salinity and temperature conditions tend to be more stable. Under these environments. Two groups of Octocorallions occur: tree-like Gorgonians and wide-bodied Alcynonaceans.

dominated by Cantonese, and four with mixed populations.

Most of the farming activities that produced terraces, now overgrown, have been abandoned. However, several villages survive by renting homes to outsiders, which became possible after the High Island Reservoir, with its new road infrastructure, was completed in 1979.

The High Island Reservoir

The High Island Reservoir cost HK$1,348 million and caused one of the largest impacts that people have imposed on the Country Parks of Hong Kong. The area was once a narrow strait between the mainland and the formerly separate High Island. Two large rock-filled dams (110 and 103 m high) were completed in 1979. These closed the strait, and sea water was pumped out. The original catchment was too small (15 km^2), thus 40 km of tunnels were built to divert fresh water to the reservoir, effectively adding 61 km^2 to the drainage basin. Today, the reservoir holds 281 million m^3 of

Accessible only by boat or a long walk, the remote fishing village of Ko Lau Wan, in northeastern Sai Kung, is one of a few remaining coastal settlements. The village supports several old Taoist shrines.

The High Island Reservoir occupies a former marine strait that flooded a series of even more ancient river valleys between about 8,000 and 6,000 years ago. This is the reason for the highly indented coastline. Ping Pai, the summit of High Island, is the tallest mountain in the distance. Shui Keng Teng forms a large island within the reservoir and is visible in the middle distance.

water, and, when at full capacity, the surface stands 61.5 m above sea level. To protect the High Island Reservoir from storm waves, 7,000 concrete dolos (an Afrikaans word for a sheep knuckle) were placed against a coffer dam that was originally used during construction (adjacent photograph).

Materials for dam construction were derived from two sites on the north side of the reservoir. Scars are still visible today. One quarry is located on the island of Shui Keng Teng and the other lies close to the Main Dam. Both were excavated into columnar-jointed tuffs (IN40, p. 136), which would not form blocks larger than 1.2 m in diameter. Because this was too small for the outer facing of the dams, granite was imported from quarries at Turret Hill, in Sha Tin. Quarrying also took place at Nam Fung Shan, on the northwestern side of Long Harbour. From there, very large quantities of tuff were extracted for use, ironically, in the Sha Tin Reclamations. This particular quarry has been landscaped by using non-native trees, giving the area a distinctive appearance.

A few small dams are also present. These block mountain streams, providing water supplies for villages. The largest straddles the Kap Man Hang in the heart of the Sai Kung East Country Park. Several large waterfalls occur along the length of this exhilarating stream, which provides one of the best wilderness walks in Hong Kong.

Dolos (inset) have been used to protect the High Island Dam. They allow water to pass between them, absorbing wave energy by converting it to frictional heat. They also shift position, further absorbing energy. Note the many holes at the back of the coffer dam, which allow water to pass through, reducing the risk of storm damage.

Columnar-jointed tuffs were quarried near the Main Dam, for use in building the cores of the High Island dams. Interestingly, the quarries are referred to as "borrow" areas, a misnomer, given that material is not returned.

The Kap Man Hang Stream provides a wild walk through the centre of Sai Kung. Many beautiful waterfalls lie along its course. Unfortunately, the route is also partially blocked by a dam at about the 100-m contour level.

The Eastern New Territories: Human Impacts

Footpaths

The Sai Kung peninsula is crossed by an intricate network of trails that originally formed links between villages or provided access to fields. Some routes clearly have the ascent of a mountain as their only reason to exist. These trails now form an integral part of the landscape but vary in their condition. Many present easy walking, but a number have been badly eroded as a result of pressure from numerous hikers, especially where slopes are steep and formed in loose materials.

Erosion is initiated by trampling, which destroys the protective vegetation and compacts the soil. Once compacted, water cannot soak into the ground, so it flows over the surface and starts to erode the soil. The path then becomes a stream channel, intercepting more water from the slopes, which, in turn, increases the volume of water flowing along the path and increases the rate of incision. A problem of accelerated erosion then develops. These effects can be minimised by using rocks to cover the path or by diverting water from the trail with oblique lines of stones.

Trail erosion is a significant problem in many areas, especially where trails ascend steep slopes, as on this section of Sharp Peak.

Golf Courses

Recently, there have been landscape changes brought about by the development of golf courses. While catering to a recreational need, they create artificial landscapes that can visually be jarring and can cause a range of environmental problems relating to water consumption and the use of pesticides, herbicides, and fertilisers.

The MacLehose Trail and other paths, have been concreted in recent decades, as along this section to Sai Wan, detracting from the natural beauty of Hong Kong.

Multiple municipal golf courses have been developed on Kai Sai Chau (below) in recent years. These form artificial landscapes set against the general background. A smaller course was also developed at Sham Chung (right), a village area not included when the Sai Kung West Country Park was originally designated.

148

LANTAU ISLAND

The geology of Lantau is distinctive because it includes a very large area of dykes. It also provides favourable conditions for slope failures on many of its steep mountain slopes. For a long time, it was considered remote and inaccessible, but, in recent decades, the degree of human landscape intervention along the north shore has escalated more than in any other part of Hong Kong.

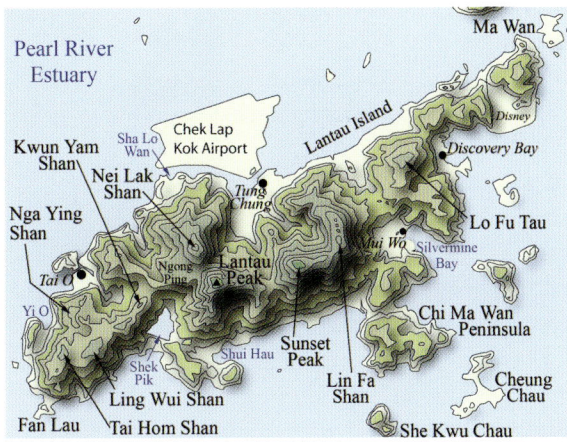

The island (146 km^2) is the largest in Hong Kong. It shares its name with Lantau Peak, which locally means Ragged Head. The mountain (934 m) is also known as Fung Wong Shan (Phoenix Mountain) and is the second highest summit in Hong Kong. The peak lies at the centre of a high massif around the Tung Chung Valley. Other mountains include: Sunset Peak, or Tai Tung Shan (869 m); Lin Fa Shan (766 m); and Nei Lak

Lantau is home to some of the highest mountains in Hong Kong (above). Lantau Peak lies in the background of the photograph below, which shows the view from Nga Ying Shan (southwest Lantau).

Shan (751 m). A number of smaller hills extend both to the southwest and northeast to form a spine of high ground along the length of the island. In the southwest, the tallest summit is Ling Wui Shan, which reaches an altitude of 490 m. In the northeast, the highest elevation is at Lo Fu Tau (465 m). Many parts of the shoreline are steep and rocky. Extensive areas of low ground only occur near the coast, at places such as Tai O, Tung Chung, Pui O, and Mui Wo.

Ma Wan Chung, photographed in 1997, was once an isolated community on the north shores of Lantau. Today it survives but is sandwiched between a bay enclosed by the Chek Lap Kok Airport and modern buildings of the rapidly expanding Tung Chung New Town.

In 1998, the population of the island was about 43,000, but rapid urban growth in Tung Chung is expected to increase this figure to a current target of 389,000. Several smaller towns also occur, each with their own distinctive character. Tai O is

The northwest Lantau coast consists of a series of muddy bays and steep hillsides. This photograph shows the view from the middle slopes of Nga Ying Shan, looking southwest over Yi O. Tai Hom Shan, on the left, rises to a height of 466 m.

Tai O (above) is located next to the muddy waters of the Pearl River Estuary and has been built over a series of tidal channels separating Fu Shan (in the background) from the rest of Lantau. Note the various ponds next to the sea, which were once used to evaporate sea water and produce salt.

the most remote settlement and is located on the northwest coast. It is particularly distinctive in retaining traditional stilted houses built over a network of tidal channels. The town was a Tanka village when the British first arrived. However, in subsequent years, Han, Hoklo, and Hakka migrants have also made homes there. Mui Wo (about 10,000 people) lies on the east coast next to a beach at the head of Silvermine Bay and consists of several small settlements and scattered village houses. The modern homes and high-rise buildings of Discovery Bay (14,500 inhabitants) form a dormitory town, situated in the northeast, that is only accessible by boat or bus.

In general, summits on Lantau are grassy, with forests occurring on the lower slopes. Most of the trees are the result of deliberate afforestation, with particularly extensive plantations occurring on the Chi

Houses at Tai O are built on stilts and many have tin walls (below). The local name for this style is pang uk, which literally means "stilted house".

Ma Wan Peninsula and around the Shek Pik Reservoir. Woodlands near the base of Lantau Peak and Sunset Peak were designated as Special Areas in January 1980. Scrub tends to be found on relatively sheltered slopes and in stream valleys. Large parts of the island are protected within Lantau Country Park South (5,640 hectares), and Lantau Country Park North (2,200 hectares), which were designated in 1978. Currently, a northward extension has been proposed to bring the remaining hilly areas within the park system. An additional marine park has been proposed for southwest Lantau, along the Fan Lau coast. The mangrove-fringed mudflats near Tung Chung have been designated as a Site of Special Scientific Interest (SSSI), as they are one of only two locations in Hong Kong where marine eel grass (*Zostera japonica*) is known to grow.

Lantau North Country Park | Proposed Country Park extension
Lantau South Country Park | Proposed marine park

Lantau Country Parks North and South protect much of the mountain landscape of Lantau. In recent years, there have been proposals to extend the northern park to offset damage done to the natural ecosystems along the coast.

Ancient Sediments, a Caldera, and Multiple Intrusions

The oldest rocks on Lantau are found along the coast northeast of Tai O. These consist of grey-to-red sandstones and siltstones that comprise the Tai O Formation. They were laid down during the early Jurassic (205–180 million years ago), at about the same time that dinosaurs roamed the earth, though no such fossils occur in Hong Kong. However, plant remains have been found in these deposits, which accumulated in river channels and on floodplains.

Most rocks on Lantau are tuffs, or intrusive rhyolite and rhyodacite dykes (IN45, p. 155), with smaller areas of granite. The oldest tuffs (164 million years old) belong to the

The Tai O Formation lies at the base of the cliff above and consists of red and grey siltstones. Note the sharp boundary with the pebbles (inset). This is an unconformity—a time gap between the rock layers, during which erosion took place. Below this surface, the rocks are 200 million years old. The pebbles above are younger (a few tens of thousands of years), well-rounded, and accumulated in river channels. The top of the cliff is made of angular colluvial blocks (IN17, p. 91) that have slid down from the hillside above.

Lantau Island: Ancient Sediments, a Caldera, and Multiple Intrusions

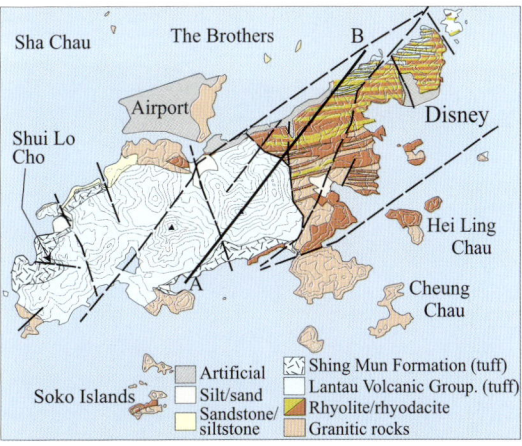

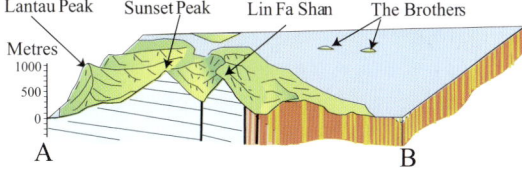

Lantau is dominated by volcanic tuffs and lavas to the southwest and by rhyolite and rhyodacite dykes to the northeast. Other parts of the island are mainly comprised of granitic rocks, with a band of sedimentary sandstones and siltstones occurring along the northwest coastline near the town of Tai O.

The Shui Lo Cho cuts deeply into volcanic tuffs of the Shing Mun Formation and is one of the most dramatic streams in Hong Kong. The water course follows an east-west fault line, with harder rocks resulting in the formation of several attractive waterfalls. Deep plunge pools cut into the softer underlying tuffs.

Shing Mun Formation (map above and adjacent photographs), which was erupted as ash during the first phase of back-arc igneous activity that affected Hong Kong (IN23, p. 105). Shortly after falling onto the land surface, some of the loose ash was washed down the volcanic slopes as lahars. These are dense mixtures of water and ash that form after rain and which are often concentrated along stream channels. Some of the volcanic materials may also have been moved down hill as a result of slope failures. A little later, about 161 million years ago, the Lantau Granite was intruded (figure, p. 107).

A second phase of volcanism occurred 18 million years later, giving rise to the Lantau Volcanic Group. These are tuffs and lavas that were produced within a NE-SW zone, bounded on each side by faults that extended from Lantau, through Sha Tin, and on to the Tolo Channel. Other rocks commonly found in Hong Kong today were also formed within this belt at about

Lantau Island: Ancient Sediments, a Caldera, and Multiple Intrusions

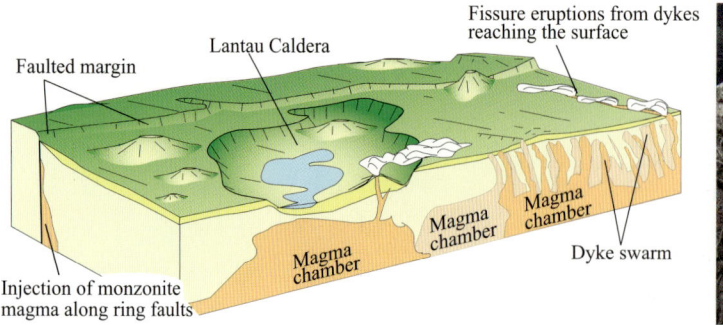

Tilting toward the southwest of the 146-million-year old rocks on Lantau has revealed a cross section through the crust at that time. The rocks were formed in a fault-bounded rift, within two broad geological settings. To the southwest, they were originally erupted as lavas (right photograph) or as volcanic ash. These materials accumulated within a depression called a caldera. To the northeast, the rocks originated as magma that cut into the earth's crust to form multiple parallel dykes. Both were probably fed by similar magma chambers at greater depth.

the same time. Examples include the Sha Tin and Needle Hill Granites. On Lantau, a fault-bounded elliptical depression, or caldera (IN40, p. 136), developed, in which tuffs and lavas of the Lantau Volcanic Group accumulated.

To the northeast, at the same time that the caldera was developing, a swarm of dykes (IN45, p. 155) penetrated the crust along an eastnortheast alignment. Two episodes of dyke intrusion occurred. The first gave rise to a set of wide dykes of rhyodacite, a rock that contains large feldspar and quartz crystals (right photograph). The second occurred a few hundred thousand years later and produced thinner dykes of rhyolite (IN45, p. 155), which has a higher silica content and more quartz.

Exposure at the surface has resulted in weathering (IN51, p. 177) of these ancient dykes. This has made evident contrasts in the resistance of the rocks and minerals to chemical decay. Joints also tend to produce lines of weakness that are preferentially weathered and eroded, sometimes into interesting forms (adjacent photograph).

Granitic rocks were then intruded, about 144 million years ago, and cooled

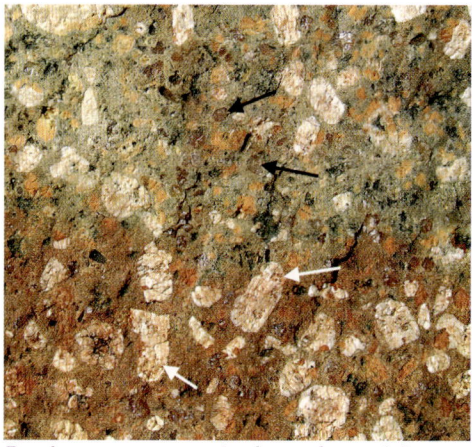

Porphyrys are common in the northeast of Lantau. These are igneous rocks that contain large crystals surrounded by a matrix of microscopic crystals. The example above is a rhyodacite porphyry, which contains large crystals of quartz (black arrows) and feldspar (white arrows). Rhyolite porphyry also occurs. This is similar but contains a little more quartz.

The rock above is a basaltic andesite dyke less than 1 m wide. Weathering of intersecting joints has generated an unusual face-like appearance.

IN45 The Lantau Dyke Swarm

The Lantau dyke swarm includes hundreds of nearly parallel ENE-trending dykes (IN04, p. 22) that make up about one third of Lantau. Several types of rock occur (rhyodacite, rhyolite, and basaltic andesite). The margins of the dykes commonly cooled too quickly for large crystals to grow. Consequently, minerals are invisible to the naked eye in what are called chilled margins. In other central parts of the dykes, which cooled more slowly, large crystals of feldspar were able to grow. These are usually surrounded by a matrix of microscopic crystals. Geologists describe these interesting rocks as feldsparphyric.

This photograph shows an example of a contact between a fine-grained rhyolite (left side) and decayed feldsparphyric rhyodacite (right side). Note the parallel bands in the rhyolite.

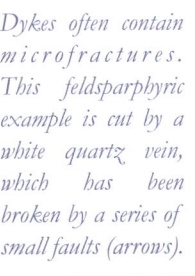

A sharp contact between two dykes is marked in the photograph. The dyke on the left is very fine-grained (inset) in contrast to the coarse-grained feldsparphyric dyke to the right.

Dykes often contain microfractures. This feldsparphyric example is cut by a white quartz vein, which has been broken by a series of small faults (arrows).

The smallest (<1 m wide) and least common dykes are dark grey, fine-grained basaltic andesites. This example cuts through a feldsparphyric rock. These dykes formed from magma that was "left over" after the rhyolites and rhyodacites had solidified.

The landscape of the northeast coast has been shaped by multiple dykes. Here, a vertical rock wall is following the side of a vertical feldsparphyric dyke.

The Lantau Volcanic Group is comprised of tuffs and lavas that underlie most of western and central Lantau and all of the area in the above photograph (taken from Kwun Yam Shan). The rocks along a major NE-SW fault have been eroded to produce the low col. The same fault continues on below the Shek Pik Reservoir (inset).

slowly within large magma chambers. Today, these rocks are found on the Chi Ma Wan Peninsula and the islands of Cheung Chau and She Kwu Chau. The last magma body to solidify (140 million years ago) produced the Fan Lau Granite, which makes up the southwestern tip of Lantau Island.

At a later stage, all of these rocks were cut by faults. Today, many Lantau valleys follow these lines of weakness. The longest example is a NE-SW fault that has determined the alignment of the northern coast and which has formed a col (mountain pass) that separates Lantau Peak and Nei Lak Shan. The same structural weakness controlled the orientation of the valley in which the Shek Pik Reservoir lies.

The Lantau Volcanic Group forms most of Sunset Peak, except for the highest 80 m below the summit, which is capped by younger tuffs.

The Chi Ma Wan Peninsula juts several kilometres southwards into the South China Sea from Lantau Island. The area consists of granite (rich in a black mineral called biotite) and is separated from the rest of Lantau by a low pass that follows a fault. The landscape is typical of a region dominated by granite, with large rounded outcrops of bare rock, as in the photograph above. Elsewhere, rounded boulders lie strewn across well-forested hillsides. Note the rock climbers on the cliff to the left for scale.

The southwest tip of Lantau is a rocky peninsula with isolated beaches. It is formed by the Fan Lau Granite and strewn with numerous rounded boulders, which are a common feature in granite areas.

Mass Movements and Landslides

Mass movements include several different processes (IN46, p. 159) by which materials are moved down a slope, either quickly or slowly, with or without water. The driving force is gravity, with particles either sliding, rolling, or flowing downhill.

The commonly used term "landslide" is not well-defined but is mostly used where slope collapse and slippage is faster and more dramatic (e.g. slumps and debris flows; IN46, p. 159). Sudden slope failures are common where there are steep inclines, heavy rains, and hot climates. Warm, humid conditions are important as they lead to the chemical decay of rocks (IN51, p. 177–8), which produces loose materials on slopes that can then move. Lantau exhibits all of these conditions, so landslides are very likely to occur.

The loose particles on a slope may also move downhill slowly, grain by grain, in a process called creep (IN46, p. 159). Although a less dramatic kind of mass movement, it is ubiquitous on most slopes with loose particles and probably transports a much greater volume of material than the rapid more impressive slope failures.

The materials that are transported down a hill by mass movements are referred to as colluvium (IN17, p. 91). On Lantau, thick colluvium occurs on lavas and tuffs. Especially thick colluvium extends down the steep hillsides from Nei Lak Shan and Sunset Peak to Tung Chung. In contrast, colluvium tends to be thinner on granites in areas such as the Chi Ma Wan Peninsula.

Many hillsides on Lantau are precipitous and are potential sites for slope collapse. In most cases, this is not a problem for people, except where steep inclines lie above built-up areas, such as at Tung Chung. This photograph shows several skyscrapers and building sites below Pok To Yan, which rises steeply to heights of over 500 m.

This example of a mass movement occurred on a small hill in northeast Lantau and is called a slump. These movements involve a block of slope material slipping suddenly down hill, leaving a curved scar where it has been detached.

Another form of mass movement involves sliding over a plane of weakness, often a joint or fault. This jointed granite shows a simple example of a slide in action.

IN46 Mass Movements

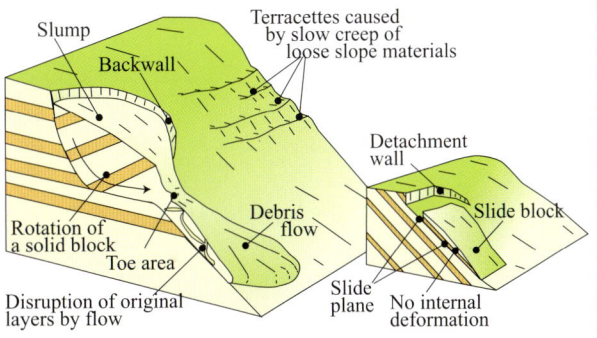

On slopes that are stable, the internal friction and cohesion of the materials resist gravitational collapse. Anything that weakens this will cause instability. Once a critical threshold is reached, a ground failure occurs. Weathering, for example, weakens rocks, producing loose materials that are prone to move downslope. Earthquake vibrations can trigger a collapse. Road cuts weaken hillsides. Infiltration of rain water fills spaces between loose particles that make up a slope and can force the grains apart, initiating a collapse.

Sometimes a spectacular collapse (slump) occurs. The material moves as a solid block, rotating around a horizontal axis and leaving a curved scar (the backwall). The lower part of the slump (the toe) is disrupted and forms hummocky terrain. Sometimes, the slump may develop into a debris flow (if wet) or debris avalanche (if drier). When this happens, the material flows like a thick fluid, and all internal structure is lost. Debris flows are common in Hong Kong and can move a great distance. Another kind of rapid mass movement is a slide, which involves a solid block moving over an inclined plane of weakness orientated in the same direction as the slope.

Slope particles may also move slowly over many years. This occurs grain by grain in a process called creep, which may lead to the formation of multiple small parallel terraces that contour along the hillside.

A slump along the catchwater to Fan Lau (above) shows a curved backwall scar, a solid block that has slipped, and a debris avalanche that has moved downhill from the slump. The debris avalanche contains loose angular boulders, gravel, and smaller materials mixed together (inset).

Many slumps have affected the Sai Kung hillside in the right photograph below. The largest has a long trail formed by a debris flow. The left photograph shows a closer view of a different debris flow, which consists of a mixture of rounded boulders and angular pebble-size rocks "floating" in a fine-grained sediment.

Coastal Landscapes

Examples of contrasting coasts on Lantau are shown to the right. Cheung Sha Beach (top photograph) is the longest in Hong Kong and lies at the head of a bay exposed to strong waves. These waves drive sand from the headlands towards the bay head, which is also supplied by sand from rivers.

Headlands are predominantly rocky (second photograph), as wave energy is focused there by wave refraction (p. 138). The example shown is a granitic area on the Chi Ma Wan Peninsula. The granite there tends to produce large curved slabs, which are defined by curved sheeting joints (IN09, p. 37).

On some beaches, there are boulders (middle photograph) that result from multiple processes. Firstly, weathering causes rock to decay on a slope. This produces a mixture of loose small particles and solid, rounded corestones (IN51, p. 177). Then a slope collapse occurs, and the material is carried to the beach as a debris flow (IN46, p. 159). Finally, the waves wash away the smaller grains, leaving boulders.

Along northern Lantau, some shores are protected from strong waves. They receive mud from the Pearl River Estuary and clay from local streams. Mangroves are common in bays such as Tai Ho Wan (fourth photograph) and support a variety of animals, including mudskippers, gastropods, crabs, and snapping shrimps. These settings are rich in marine life. On average, 116 animals occur per square metre—about 66 crustaceans and 50 molluscs.

Artificial shores (bottom photograph) are a new feature dominating northeast Lantau. These coasts are made of large granite boulders that form a sterile and unattractive landscape.

Lantau Island: Coastal Landscapes

Tidal flats (IN47, p. 162) are broad, gently sloping areas that are exposed and flooded by tides. A good example occurs at Shui Hau (photograph above and map) on the southwest coast of Lantau. The tidal flat occupies a valley that formed an estuary when first flooded some 6,000 years ago. Since then, river sediment has infilled the bay. A sand bar has built up across the valley and now forms a beach that separates the clayey paddy fields from the muddy sands.

The tidal flat has an interesting surface. Large areas are covered with angular pebbles (covered with oysters) and coarse-grained sands. These are former river deposits. Elsewhere, the sediments are finer-grained and have a wave-formed, rippled surface (IN48, p. 163). The ripples vary in orientation and distance between each other, reflecting differences in wave directions and tidal flows.

The two main beaches at Shui Hau have different features. At the head of the tidal flat, the sand is coarse and uniform. Beach sand to the northeast (adjacent map) is quite different

The Shui Hau tidal flat (from the northeast) is flooded in the photograph above but can be recognised by the swirls of muddy water in the right half of the bay. Note the beach to the extreme right. This is part of a sand bar (marked on the map) that separates former paddy fields from the modern coast. The rounded hill on the far shore is made of granite, with the lower ridge to its right consisting of an unusual rock called quartz monzonite (see rock composition figure, p. 86). The foreground is dominated by tuffs. The Soko Islands lie in the middle distance. Islands further away are within mainland waters.

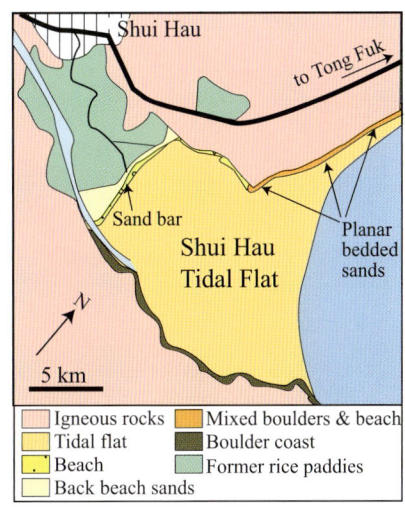

161

IN47 Tidal Flats

Tidal flats are comprised of three main environments: supratidal, intertidal, and subtidal. Supratidal flats occur above the mean high tide and are flooded only during the highest water levels and, occasionally, during storms. These settings are sensitive to climate. They tend to develop as salt marshes in temperate regions and mangrove stands in tropical areas.

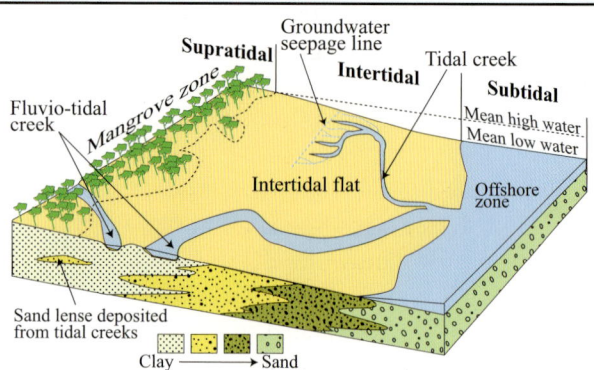

The figure above shows the division of tidal flats into three major zones and the tendency for sediment grain size to become coarser seawards.

Intertidal flats occur between the high and low tide levels and alternately are submerged and exposed. Sediments vary in Hong Kong, with both mud and sand accumulating. Mangroves and grasses are absent. Wave energy is highest at the seaward end of an intertidal flat, and sand tends to be deposited there. As waves cross the gentle slope, energy is lost through friction, and mud can begin to settle. Consequently, grain size tends to decrease landwards. Tidal creeks are formed by rivers that cross the flat during low tide and by streams fed by groundwater seepage on the flat. Wave-formed ripples and hummocky surfaces are often present. The latter are caused by burrowing crabs and people collecting clams.

Very gentle flats, in sheltered areas, are associated with fine-grained mud. This is the case at Mai Po (upper). Tidal flats exposed to more energetic waves and/or faster tidal currents, tend to be steeper and sandier, as at Shui Hau on Lantau (lower).

The sub-tidal environment lies below the mean low-water level and is permanently submerged. It is characterised by relatively coarse-grained sediment, off-shore sand bars, and tidal-scoured channels. Generally, the slope of the subtidal zone is steeper than that of the intertidal flat.

(adjacent photograph), consisting of alternating thin layers of black and light-coloured sands. The layers are nearly horizontal. Geologists call this kind of layering planar cross bedding, and it is a common feature on beaches. The layering has developed because there are two differing sets of minerals, with contrasting densities, at this location. These are relatively heavy, black hornblende as well as light-weight, light-coloured, quartz and feldspar. These minerals were separated from each other by wave action (swash and backwash; IN42, p. 141), which, over time, has also caused the two sand types to be stacked on top of each other. Source rocks are important here. The sand at the first beach lacks a supply of hornblende (derived from tuffs and quartz monzonite), so layering is not visible there.

The photograph shows alternating black- and light-coloured sand layers, with an erosion surface between the arrows. Note how layers are cut out at the erosion surface. Waves deposited the sediment and then washed some of it away to form the erosion surface. Further sand layers were then laid down.

A third kind of coastal deposit occurs near Tong Fuk. There, a delta has formed, fed by a river. Pebbles delivered by the stream have been reworked at the coast by powerful waves to form a 10 m high pebble beach.

The Tong Fuk pebble beach rises 10 m above the sea.

IN48 Current Ripples

Current ripples are formed when water flows over loose sand. They do not occur in mud or very coarse sediments; they have an asymmetrical profile. Symmetrical types can develop if there is a to-and-fro motion of the water. The distance between the crests of adjacent ripples is called the wavelength (A). The steep part is called the lee side, with the gentle stoss slope facing upstream.

The photograph shows well-developed straight to catenary ripples on the Shui Hau tidal flat.

Flowing water erodes particles from the gentle stoss side, which lowers this surface (B). The grains are gradually washed up the gentle slope, until they fall down the steep lee side and are deposited. This continuing process results in the ripple migrating downstream, leaving an asymmetrical ripple with thin inclined layers (C). Viewed from above, ripples may be straight, catenary, linguoid, or lunate in shape (D). The particular type that develops is related to the speed of the water flow, grain size, and water depth.

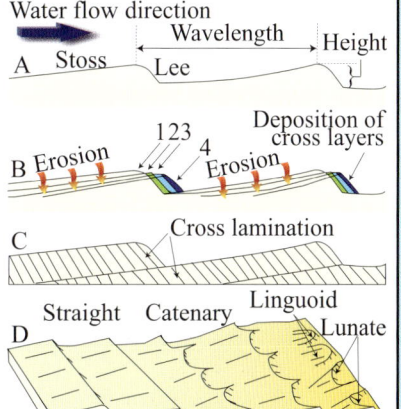

Human Impacts

Artefacts found on Lantau suggest people have lived there for thousands of years. A small stone circle at Fan Lau probably dates to the Neolithic (10,000–5,000 years ago). Rock carvings at Shek Pik are believed to be Bronze Age (about the second millenium BC). However, human impacts on landscapes were probably localised and of a small scale. As time passed, populations expanded, and Chinese emperors and others took greater interest in the area, necessitating a military presence.

Forts

Piracy was a major problem throughout southern China for several centuries (IN49, p.165). In the face of maritime weakness and the lack of a significant navy, the Chinese built forts for coastal defence. Landscape analysis was a vital element in the siting of these forts. The locations had to be strong defensively and offer wide views of the sea. Thus many, though not all, were built above coastal cliffs. Fan Lau Fort was built in the 1720s on the southwestern tip of Lantau. On the northern slopes of Shek She Shan in the Tung Chung Valley, a battery was constructed at Shek She (1817), which possessed two cannon emplacements and seven guard houses. Adjacent, to the south, was Tung Chung Hau Shuen, built in 1817, with eight guard houses and a garrison of 30 soldiers. There is little record of the fort after 1877. The Tung Chung Walled City (Tung Chung Fort) was constructed in 1832, with four rubble-filled walls enclosing an area 69 m long and 81 m wide. The front wall is almost 5 m thick. The fort held six cannon and two guard houses.

Tung Chung Fort has three gateways: Main Gate, East Gate, and West Gate. It was converted to a police station by the British in 1898 and to the Anglo-Chinese School during World War II. More recently, it has served as an office and a primary school.

Tai Yu Shan Fort (Fan Lau Fort) is located on the southwestern tip of Lantau Island, 116 m above the sea. Built in 1720–23, it comprises four rubble-filled walls and an entrance facing to the east. There were eight cannon emplacements and 20 guard houses. British officers found the fort unmanned in 1842.

IN49 Pirates and Forts

Given China's lack of a navy, pirates were dealt with by building a chain of coastal forts from the Shandong Peninsula to Hainan, with Hong Kong as an important link. By the Tang Dynasty (618–907 AD), garrisons were established at many locations, with the earliest recorded place-name being Tuen Mun, meaning "garrisoned entrance", where 2,000 soldiers were stationed. The garrison remained unchanged through the Song Dynasty (960–1279 AD). Records of piracy increased during the succeeding Yuan Dynasty (1279–1368).

Early in the Ming Dynasty (1368–1644), the Emperor imposed a ban on going out to sea. In 1384, two military districts were set up in Tuen Mun to control the western and eastern parts of Hong Kong. The Portuguese arrived at Tuen Mun Bay from Malacca in 1514 and settled there, building forts before they were driven out to Macau by the Ming forces. Around 1536, six guard stations were set up between Long Pak (south of Macau) and Mirs Bay. Three of the stations were in Hong Kong. Naval forces became even weaker during the Ming period, and piracy increased.

During the Qing Dynasty (1644–1911), pirate groups increased from a few thousand men to bands of 50,000. During the 1650s, the pro-Ming Cheng family controlled the Fukien coast. By the 1730s, Cheng Lin-cheung was based above Lei Yue Mun, and his reputation as a ferocious fighter led to the naming of Devil's Peak. In 1661, the Qing government ordered the evacuation of all areas within 50 li (29 km) of the coast to prevent supplies reaching Ming loyalists. All coastal forts were destroyed and

Excavations within the grounds of Tung Lung Fort revealed a settlement and yielded many artefacts.

stronger defences built inland. Military observation posts were established at Lion Rock, Ma Tseuk Leng, Tai Po Tau, Tuen Mun, Maau Chau, Shing Shan, and Fat Tong Mun. The evacuation was abandoned in 1669. When people returned, piracy was still rife and walled villages were constructed for protection.

Subsequently, seven forts were set up in Hong Kong at Tuen Mun, Kowloon, Shing Shan, Tai Kwan Ying, Fat Tong Mun, Tai Po Tau and Ma Tseuk Ling, each occupied by 10–50 soldiers. Between 1723 and 1735, the Fat Tong Mun and Tai Yu Shan Forts were built to combat increasing restlessness. In the eighteenth and nineteenth centuries, forts were built at Wang Chau, Kwun Chung, Tsiu Keng, and Ma Tseuk Ling. The Tai Pang Battalion was established to protect the coast, garrisoned in Kowloon Fort, Kowloon Hoi Hau, Tai Yu Shan, Tung Chung Hau, and Hung Heung Lo (on Hong Kong Island). A battalion of 482 men was stationed at Tung Chung Walled City (built in 1832). With the outbreak of the Opium Wars in 1840, defences were strengthened, and the Tsim Sha Tsui Fort (130 men) and the Kwun Chung Fort (75 men) were added. The Opium Wars ended in 1842, and the last two large pirate bands were finally beaten in 1850 by British and Manchu forces.

Salt Pans

A distinctive feature of Tai O is the presence of numerous shallow ponds landwards of a new typhoon shelter. These are the remnants of a once thriving salt manufacturing industry that dates back to the Tang and Sung Dynasties (seventh to thirteenth centuries), when salt production thrived along much of the low-lying parts of the coast from Tung Chung to Tai O. Salt commissioners collected taxes on the salt.

Shallow ponds at Tai O were once salt pans, with hard floors surrounded by embankments. Sea water was allowed to enter at high tide and then the salt pan was sealed off, leaving up to 15 cm of standing water to evaporate. Excess sea water was drained away on the third or fourth day, leaving 2-3 cm of water that was allowed to dry. Salt was collected on the fifth day.

Many of the salt pans were abandoned during the coastal evacuation period (IN49, p. 165). In the 1720s, tax concessions were offered for those who converted salt pans and marshes to paddy fields. Limited salt production lasted until the 1960s, after which the industry died out.

Tropical Hardwoods

Hong Kong has imported tropical hardwoods for many years, and Yam O in northwest Lantau has served as an important log pond for temporary storage. Log rafts were once towed from here to saw mills in Tsuen Wan. In recent years, the area has been partly reclaimed for road and rail infrastructure. Yam means "dark", a name considered unsuitable for the link station to Disneyland that was built there. It is now renamed as the brighter sounding Sunny Bay.

Tropical hardwoods from SE Asia were once stored at Yam O (above, in 2006). The photo below shows a single log storage raft in 1994.

Mining on and near Lantau

There are no active mines on Lantau today, but it was once a source of lead, silver, tungsten, and graphite. Other uneconomic minerals contain zinc, tin, fluorine, and beryllium. Silver Mine Bay, for example, was named after a mine 1 km northwest of Mui Wo that was active until 1896 and produced galena (yielding lead) and small

amounts of silver. The ore was extracted using a tunnel (or adit) that followed a fault that separated a rhyolite dyke (IN45, p. 155) from granite. All that can be seen today is the adit entrance, now walled up.

Wolframite (IN21, p. 100) was mined at Sha Lo Wan on the north coast and yielded tungsten. In the late 1950s, the Far Eastern Prospecting and Development Corporation started large scale workings, but operations ceased in the 1960s. Disused mine facilities remain about 1 km east of Sha Lo Wan village.

Graphite was mined by the Ng Fuk Black Lead Mining Co. Ltd. on West Brother Island, 4 km north of Lantau, from late 1952 (IN58, p. 216). Production stood at 3,500 tonnes a year in the 1960s with operations ceasing in 1971. Mining occurred entirely below ground to depths of 90 m below sea level. There are no remains of the workings today because the island was levelled as part of the international airport development.

From 1939 until recently, kaolin (a white clay used for pottery) was mined at several places in Hong Kong. Total production was 6,663 tonnes in 1960, and, during the 1960s, 5,000 tons were exported to Japan annually. Kaolin, derived from weathered feldspar in granite, was mined at Miu Wan (northeast Chek Lap Kok Island). A second mine was in use between 1981 and 1988. Chek Lap Kok also provided granite building stone up until World War II. The Chi Ma Wan Peninsula was similarly exploited for granite building-stones.

There were once three entrances to the silver mine near Mui Wo. Two have been blocked by earth and rubble and can no longer be seen. This one is bricked up about 20 m into the cave. The old mine is now flooded with a 2 m deep river.

Religion and Lantau

Most towns and villages have their own temples, but for many decades Lantau has also served as a secluded island and mountain retreat for Buddhist monasteries. The largest centre of religious activity is on the Ngong Ping Plateau, which lies at a height of 400–500 m just below Lantau Peak and Nei Lak Shan (adjacent photograph). A shrine to Buddha was built

Ngong Ping Plateau lies to the left in the photograph below. The giant Tian Tan Buddha sits on Muk Yue Shan and is just visible on the skyline. Numerous monasteries are located on the plateau, with others occurring to the west and extending down towards Tai O. The high mountain summit is Lantau Peak, rising to a height of 934 m.

there by three monks in 1920, which became the Po Lin Monastery in 1924. This has become a significant centre following intermittent expansion since it was founded. In 1993, the 26-m-high Tian Tan Buddha, clad with copper and sitting on an 8 m high pedestal, was completed on Muk Yue Shan.

Other monasteries are present on the plateau and the Keung Shan and Man Cheung Po areas. A Trappist Haven Monastery is located at Tai Shui Hang in northeastern Lantau (near Peng Chau). This religious establishment is home to Roman Catholic monks and adopted the name Our Lady of Joy Abbey on 15 January 2000.

The Tian Tan Buddha sits on a low peak on the Ngong Ping Plateau. The statue dominates the landscape for many miles around. A new cable car system, from Tung Chung to Ngong Ping, was opened in 2006 in order to help transport the many thousands of pilgrims and tourists who regularly visit the site.

A rather pleasant human impact is located near the main path from Tai O to Tsz Hing Monastery. This is the Lung Tsai Ng Yuen Chinese garden (designated a Site of Special Scientific Interest). The fish pond is covered with lilies, and the area is surrounded by wild countryside. It was built and planted by the late Mr. Woo Quen-sung.

Expanding Infrastructure

Chek Lap Kok Airport, and the associated infrastructure, is the most dramatic landscape transformation deliberately carried out in Hong Kong. This vast project flattened four islands (Chek Lap Kok, Lam Chau, West Brother Island, and East Brother Island), translocated vast quantities of marine mud, and replaced it with sand and rock to form a new island.

Opened on 6 July 1998, the entire Airport Core Programme (ACP) took 6 years to build and cost HK$155 billion to implement, the airport itself costing HK$70.2 billion. The ACP comprised ten infrastructure projects. These included 34 km of roads, tunnels and bridges, an express rail-link, major land reclamation projects on Hong Kong Island and in Kowloon, a third cross-harbour tunnel linking Hong Kong Island and Kowloon, and a New Town at Tung Chung. The Tsing Ma Bridge is 2,200 m long, weighs 55,000 tonnes and is the heaviest and longest road and rail suspension bridge in the world. According to the Guiness Book of Records, the project was the world's most expensive airport development.

The airport platform extends over 1,248 hectares, which is almost four times the size of the old airport at Kai Tak, and was constructed on and between the two islands of Chek Lap Kok and Lam Chau, which were levelled to form 25% of the platform area. Platform formation

The North Lantau Expressway was built to provide access to the new airport at Chek Lap Kok. The section above runs along the northeastern coast of Lantau, passing Yam O Wan on the left. The link through the hills is a branch route to the new Disneyland resort (below), passing Constellation Lake in the extreme right. The resort was built as a separate project after the link to Lantau had been constructed—an example of knock-on developments when a new area is opened up. It entailed a large reclamation project that filled in the former Penny's Bay (Chok Ko Wan).

The 2.2-km-long Tsing Ma Bridge was built as an essential link to the new airport. Its towers rise to 206 m, and there is a 62-m clearance for shipping. A second bridge, the 1,670-m-long Kap Shui Man Bridge, completes the link via the island of Ma Wan. This also holds the record for the longest cable-stayed bridge that supports both vehicle and rail transportation.

initially comprised the dredging and removal of 69 million m³ of mud, which were transported to the South Cheung Chau Marine Disposal Ground. Levelling of Chek Lap Kok and Lam Chau produced 108 million m³ of soil and rock. A further 7 million m³ of soil and rock were obtained by levelling the East and West Brothers Islands, with another 7 million m³ of material being sourced from other places. In addition, 76 million m³ of dredged marine sand were brought to the site. Exposing the buried old river sand at the seabed required dredging about 30 million m³ of soft muddy overburden.

In total, the contractors had to dredge and re-handle more than 238 million m³ of material, which was accomplished in the short time of two and a half years. By way of comparison, dredging 238 million m³ is equivalent to picking up and carrying away nearly two million Hong Kong double-decker buses.

Importantly, prior to any work commencing, accurate marine surveying and ground investigation were required to determine the dredging level over the site. This surface largely coincided with the top of the sandy Chek Lap Kok Formation and the base of soft marine mud that had accumulated over the last few thousand years. This level was very irregular and varied from 0 to 28 m below Principal Datum (about modern sea level). The irregularities in the surface of the Chek Lap Kok Formation generally corresponded with the locations of ancient river channels (map above), most notably a large channel that extended northwards from the Tung Chung Valley. Engineers calculated that the removal of

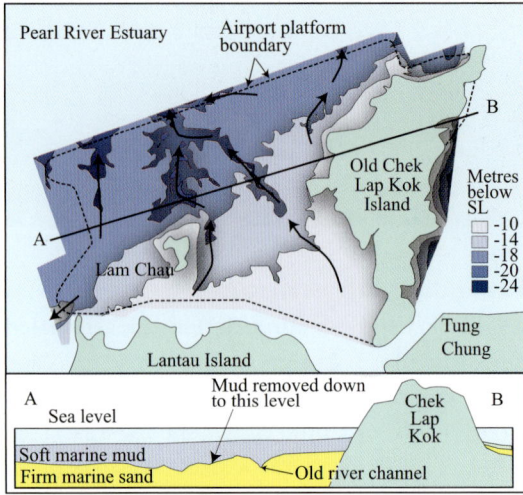

The map shows the old islands of Chek Lap Kok and Lam Chau as they used to be, with the airport platform shown by the dashed line. The sea and sea floor mud have been cut away to show the topography as it was about 18,000 years ago. Note the river channels that once formed the surface here. These were infilled by soft mud when the sea rose and flooded the area. The mud can be seen resting on top of the older surface in the cross section. This is the material that had to be dredged away to provide a stable foundation upon which to build the airport platform.

Reclamation work being carried out in 1994 was of an enormous scale. Rock from the original two islands was used as armour, but large quantities of imported sand were also needed to infill the platform area.

Lantau Island: Expanding Infrastructure

Airport construction was well under way by November 1996 (upper photograph), developing on a platform that had been constructed in only 2.5 years. The lower photograph shows the airport in 2006, several years after completion.

each 10-cm layer of material from over the entire site required an extra 1 million m³ of fill to replace it, in addition to the removal and disposal of the 1 million m³ of overburden. Consequently, an accurate model of the past geography of the site, obtained from geophysical surveys and boreholes, was essential for sound construction and cost-effectiveness.

Marine Reserves and Dolphins

The Sha Chau and Lung Kwu Chau Marine Park was designated on 22 November 1996, covering a total sea area of about 1,200 hectares. The park helps to protect fish of the Engrulidae, Scieanidae and Clupeidae families, which are important food sources for the local pink dolphins.

The Chinese white dolphin (*Sousa chinensis*) is confined to the brackish estuarine waters to the north and southwest of Lantau Island, though they sometimes are seen just east of Lantau as well. This magnificent animal was still common in 2007, but under significant environmental pressure (IN50, p. 172) as a result of increasing human activities. Also known as the Indo-Pacific hump-backed dolphin, they are born dark grey and become mottled grey to white before turning pink in adulthood. They are most commonly seen following the nets of trawling vessels.

Pink dolphins (lower photograph) are commonly seen in the waters of western Hong Kong. A marine reserve around the Sha Chau and Kwu Ling Chau islands (upper photograph) was established to protect these animals.

IN50 Dolphin Habitat Problems

Estimates of dolphin numbers vary considerably. One report indicates 150 dolphins in Hong Kong waters, with another 200 in mainland seas. Other estimates suggest there are about 1,000 in the Pearl River Estuary, with 450 in Hong Kong. These estimates indicate that numbers drop to about 130 in the winter. Whatever the true situation, the dolphin habitat has been severely degraded in recent decades.

Large tracts of the shallow-water breeding grounds for the fish upon which they live have been lost to reclamations in the last 10 years. A number of seabed pits have been excavated north of the Chek Lap Kok Airport and used to store toxic mud from dredging operations in and near Victoria Harbour. Several sewage outfalls also discharge into the area (map).

Dolphins that survive these pressures still have to contend with high-speed boats passing through the designated shipping lanes. An aviation fuel depot for the airport has also been built on Sha Chau, within the marine reserve, although the islands are excluded. Several parts of the sea floor have been dredged for sand in recent years, and it has been proposed that these sites be used to dump uncontaminated mud.

Several new bridges have been proposed that would link Lantau with Macau, Tuen Mun, or the north coast of the western waters. All would inevitably further disturb the dolphin habitat.

Young dolphins, such as in this group, tend to be mottled grey rather than pink.

Urban pressures from towns, such as Tung Chung and Tuen Mun, and the growth of infrastructure along the north shore of Lantau are having a profound effect on the waters that constitute the home to pink dolphins. This photograph shows the airport platform with Nei Lak Shan rising to 751 m in the background. Immediately below the viewpoint are toxic mud pits.

During the last decade, human activities have increased dramatically in, and adjacent to, the western waters of Hong Kong. The map shows some of these impacts, including toxic mud pits, reclamations, sewage outfalls, and infrastructure developments.

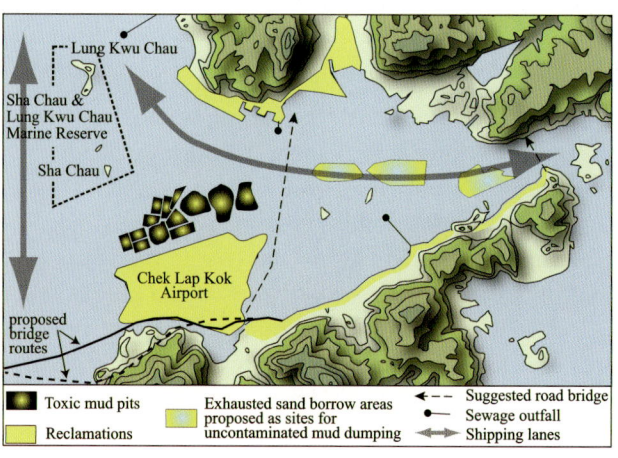

KOWLOON AND THE LION ROCK RIDGE

Kowloon is the largest urbanised region in Hong Kong, which, together with the northshore of Hong Kong Island, forms the centre of metropolitan Hong Kong. These two areas demonstrate, perhaps more than most, the critical importance to cities of understanding the natural world. Slope failures threaten people and structures. Foundation construction requires a thorough understanding of ground conditions, particularly of rock weathering. Consequently, a knowledge of the geology below an urban area is vital for sustainable city developments.

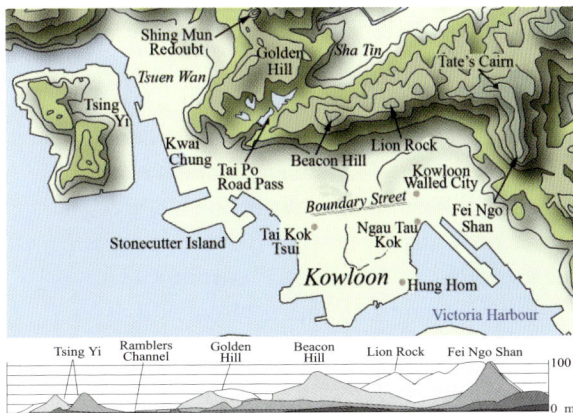

Kowloon is surrounded by water to the south and west and by mountains to the north and east. The northern hills form an east-west ridge that separates Kowloon from Sha Tin. There are several prominent peaks, including: Beacon Hill (458 m), Lion Rock (495 m), Tate's Cairn (577 m), and Fei Ngo Shan (602 m). At the old Tai Po Rd. Pass, this mountainous ridge turns north and serves as a water gathering ground for four reservoirs constructed in the valley.

The Kowloon Peninsula is a low-lying promontory of granitic land that extends southwards from the Lion Rock range of hills into Victoria Harbour. Extensive modification of the original topography has resulted from coastal reclamation, the lowering or removal of many natural hillsides, and the filling of former valleys.

The Kowloon Peninsula is a low-lying area with several small hills. It is bounded to the north and east by mountains up to 600 m, and to the south and west by Victoria Harbour and the seas of the western anchorage. Boundary St. once formed the northern border of British Hong Kong.

Feng Shui traditions suggest that a region lying to the north of water and to the south of hills, as does Kowloon, has Yang (male) characteristics. Areas lying to the south of water and to the north of hills, as does Central, are said to be Yin (female). The two together display excellent Feng Shui, which is enhanced by the way in which the Kowloon Peninsula lies within the arms of the north shore of Hong Kong Island.

Today, Kowloon is characterised by numerous artificial slopes, and by valleys that have been infilled to form development platforms.

The modern coastline of Kowloon is entirely artificial, having been subjected to several phases of reclamation since 1860. The urbanised area extends northwards along the western coastline to the main container port of Kwai Chung and the town of Tsuen Wan. The island of Tsing Yi, in the extreme northwest, serves today as a residential dormitory and as an infrastructure hub linking Kowloon, Tsuen Wan, Lantau, and the western New Territories.

The first treaty with China, signed on 26 January 1841, ceded Hong Kong Island to Britain in perpetuity, but this did not include the Kowloon area. A second treaty, in 1861, included Stonecutter's Island and Kowloon south of Boundary St. (adjacent history). Several decades later, in 1898, Kowloon north of Boundary St. and the rest of the New Territories were obtained on a 99-year lease.

The British had considered the acquisition of Kowloon essential for military reasons in order to enable the new rulers to control both sides of Victoria Harbour. At the same time,

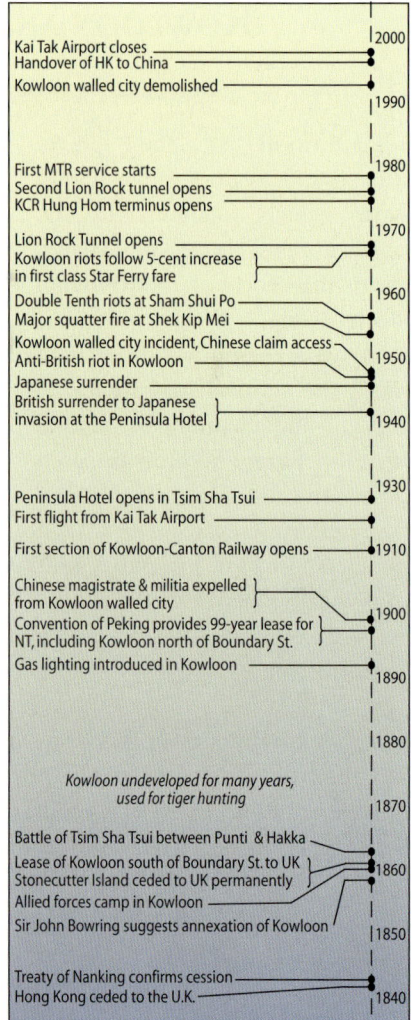

Historical summary of major events and the development of Kowloon since 1840.

The Kowloon walled city continued to be administered and claimed by China, even after the treaties that ceded the Kowloon Peninsula to the British were signed. Lion Rock rises in the background.

Kowloon and the Lion Rock Ridge: Geology and Weathering

After many years with few controls and a unique quasi-autonomous status, the Kowloon walled city had become a squalid high-density settlement with few rules. The photograph above shows the area on 1 May 1991, prior to demolition in 1994. It has since become an urban park.

The rocks of this region are dominated by several types of granite intruded as separate magma bodies (map below). The largest is the Kowloon Granite, a nearly circular pluton that underlies Kowloon and northern Hong Kong Island.

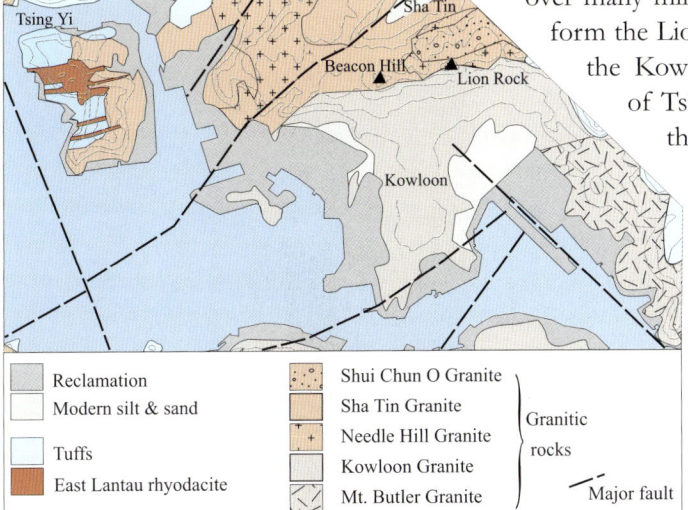

Chinese interest in the area increased with the arrival of the British, and between 1843 and 1847, a walled city was constructed on a site first fortified in 1668. This housed a garrison of 250 soldiers, which was later raised to 500 in 1898. After the British occupation of Kowloon, Chinese officials continued to occupy the walled city, but in 1899 the British drove out the Chinese military. Nevertheless, it retained its status as a Chinese enclave and gradually developed into a semi-lawless and squalid, network of narrow streets and tall buildings. With Chinese agreement, the buildings were finally demolished in 1994, and the area was turned into a park.

In 1861, British-administered Kowloon supported a population of 3,000, largely in coastal fishing villages, and a few rice-farming communities. By 1887, the population had grown to 15,000. Following the 1898 lease, the population was 43,000 in 1901, 1,250,000 in 1959, and 2,040,000 (about 30% of the total population of Hong Kong) in 2004.

Geology and Weathering

This region is dominated by several suites of granitic rocks that formed independently over many millions of years. The oldest form the Lion Rock Ridge and underlie the Kowloon Reservoirs and parts of Tsing Yi. These rocks include the Needle Hill Granite and Sha Tin Granite, which originally solidified from hot liquid magma about 146 million years ago. The emplacement of these igneous bodies was followed by yet another large magma intrusion that gave rise to the Shui Chuen O Granite

Urban Kowloon and the Lion Rock Ridge (right) are both underlain by granites, though these formed at different times. The Kowloon Granite extends below the urban area and is highly weathered.

(144 million years old). The final phase of activity brought about the formation of the Kowloon Granite and Mt. Butler Granite about 140 million years ago.

The Kowloon Granite extends below urban Kowloon and northern Hong Kong Island. It originated as a large mass of magma in a pluton (IN04, p. 22). After slowly cooling over several hundred thousand years, it formed a medium-grained granite. To the east and south, the granite is bordered by older volcanic rocks that formed the original roof rocks (ones that lay on top) of the pluton. In general, the granite is uniform in its composition and texture, although near the centre of the intrusion, around King's Park, the rocks are porphyritic, which means they contain large feldspar crystals surrounded by other smaller crystals. The manner in which these rocks break down through weathering (IN18, p. 92; IN51, p. 177) is important for foundation design. For example, where thick weathered layers occur, piling has to penetrate deeper to reach a solid base.

Weathering Characteristics of the Kowloon Granite

Thick layers of weathered rock are typical of tropical and sub-tropical landscapes, especially on granite (IN51). Hong Kong is unusual in the abundance

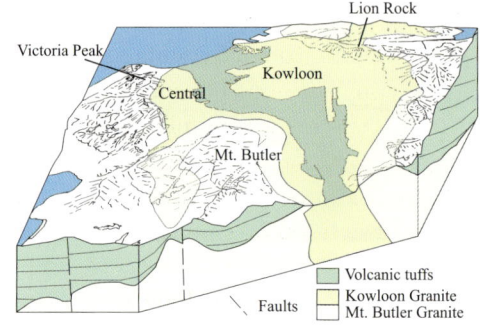

The Kowloon pluton is a sub-circular outcrop of granite that originally cooled and solidified from a large body of magma that was intruded into the earth's crust. Today, it provides the bedrock foundation for Kowloon and northern Hong Kong Island.

Kowloon is underlain by a medium-grained biotite granite (above). This is composed of grey-to-pink feldspars, greasy grey quartz, and black biotite. The granite is unusual in showing a high degree of uniformity across its entire outcrop area.

IN51 Weathered Rock Layers

All rocks at the surface of the earth are subject to attack by the atmosphere and, as a result, decay to varying degrees. Weathering is controlled by several factors, the most important of which are the climate and the rock type. Climate determines the amount and distribution of moisture and temperature, and the rate and degree of chemical reactions. Hong Kong experiences a sub-tropical monsoonal climate that is characterised by high temperatures and heavy, seasonally distributed rainfall. Intense weathering has occurred continuously in Hong Kong for several million years. As a result, thick layers of weathered rock, up to 100 m or more, have developed, especially on top of granites.

Granite plutons in Hong Kong once lay about 2 km below the then ground surface. Subsequent erosion of the rocks covering them has exposed the granites to different environmental conditions. Removal of the overlying layers allows granite to expand, and cracks, called sheeting joints (IN09, p. 37) to develop parallel to the upper surface of the pluton. In turn, these produce zones of weakness in the rock mass, which facilitates weathering.

Granitic rocks in Hong Kong mainly contain quartz and feldspars, with some biotite (IN21, p. 100). Weathering decomposes the feldspar and biotite into clay and the quartz into sand. Local drainage and groundwater conditions control the moisture content of the rocks, and this affects the nature and the rate of weathering. In particular, moisture levels influence both the amount of clay

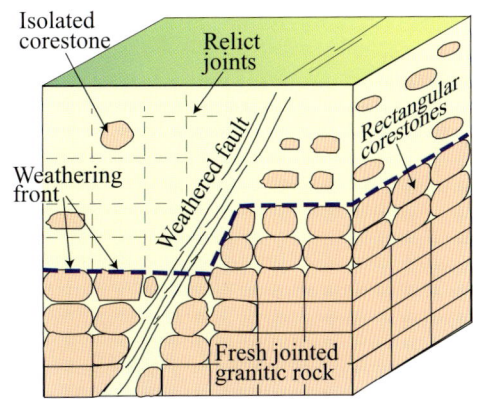

Weathered profiles involve a transition from fresh to decomposed rock. This may be gradual or abrupt, with a well-defined boundary, called the weathering front, occurring between the fresh and decayed materials. Weathering proceeds along lines of weakness, such as joints and faults, because these provide access for water. Consequently, the shape of the weathering front surface reflects these structural controls, producing sudden steps and irregularities in the profile. Furthermore, as weathering intensifies in any particular part of the profile, fresh rock becomes more and more isolated as corestones. Eventually, the entire rock body may become decomposed.

and the type of clay that occurs in the weathered material.

Thick layers of weathered rock are complex, comprising a matrix of clayey and sandy material that surrounds corestones. The latter are relatively unweathered, often rounded blocks of rock that can be up to several metres in diameter.

For a particular rock, the character of the weathered material is determined by the both the joint pattern and the spacing of joints (IN09, p. 37) in the original rock. These influence the ingress of water. The shape of the ground also plays an important part. In general,

continued on next page......

IN51 continued

upper slopes are well-drained and lower slopes are less well-drained, with valley floors commonly being characterised by impeded drainage.

At the bottom of the weathered rock layer there is usually a well-defined boundary, the weathering front (figure, p. 177), between the weathered materials and the less-weathered rock below. However, weathering along joints commonly extends for several tens of metres below the weathering front. The location and characteristics of the weathering front are important for both slope and foundation design and for tunnelling projects.

The lower part of the weathered rock layer is generally undisturbed and, importantly, preserves the original rock and structures, such as joints and quartz veins. In contrast, the upper part of the profile commonly shows evidence of disturbance by gravitational creep (IN46, p. 159), settling, and bioturbation (mixing of the material by organisms such as worms, termites, and plant roots). In Hong Kong, this mobile zone is usually less than a metre thick.

Weathered profiles vary in the density of corestones that they contain. The upper photograph shows scattered corestones in one area, but not in others. The lower photograph shows a profile that is entirely corestone-free.

The photograph to the right shows a cutslope with a very irregular weathering front. Fresh rock occurs to the right next to completely decomposed granite (on the left). Note in particlar the sharp vertical boundary between these two materials. Irregularities like these often affect construction costs in Hong Kong. A building placed over thickly weathered materials, such as those on the left, would need deep pilings to reach a solid foundation, yet a few metres to one side there might be solid rock close to the surface.

of engineering borehole data, which has allowed the topography of the weathering front, the boundary between fresh and weathered rock to be mapped (adjacent figure), and the distribution of corestones to be investigated.

Below Kowloon, the buried surface of the weathering front exhibits a gently undulating topography consisting of enclosed, basin-like depressions between dome-shaped rises. A rectilinear, structurally-controlled pattern can be discerned, which is dominated by NE-trending ridges and valleys, with subsidiary SE-trending valleys. This structural pattern was evident in the rather angular shape of the original coastline (upper map, p. 182). Similarly, a clear relationship exists between low points in the weathering front and the main mapped faults (adjacent figure).

The weathering front ranges in altitude from a high of 231.4 m above sea level at Tai Wo Ping, below Beacon Hill, to a minimum of 101.3 m below sea level at the northern edge of Sham Shui Po, a total range of 333 m. Overlying the weathering front is weathered material that ranges in thickness from less than 1 m below Beacon Hill to about 92 m immediately to the south of the Beacon Hill Ridge. Over most of the peninsula, the weathered rocks are between 10 to 20 m thick, but are generally thinner on hillsides.

Corestones (IN51, p. 177) occur within the weathered sequences. Their distribution is important for engineering projects. They are not always present, occurring in only about 25% of the boreholes drilled in Kowloon, and usually are found as clusters, commonly below ridges or high ground. Their occurrence appears to be related to major joint patterns and spacing and to the local ground water conditions.

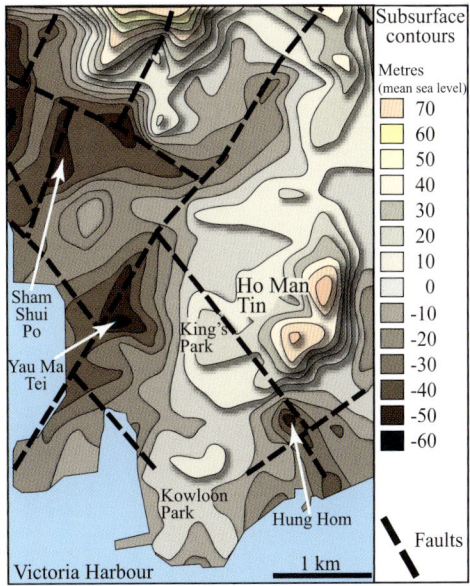

The map shows the surface of the weathering front (or fresh rock) that lies below the modern ground surface in Kowloon. Note how irregular this surface is. Beneath Ho Man Tin Hill, the weathering front forms a large dome near the centre of Kowloon, which rises to 50 m above mean sea level. Low ridges extend outwards from Ho Man Tin, one under King's Park Hill to the west and one under Kowloon Park to the southwest. The lowest points occur below Hung Hom in the southeast and below Yau Ma Tei and Sham Shui Po in the west, where it descends to 60 m below mean sea level.

The photograph shows a cutslope in Ho Man Tin that exposes a deeply weathered profile. Note the variable concentration of the protruding corestones of fresh rock surrounded by friable weathered materials. The corestones are arranged in a rectilinear (right angle) pattern, which reflects the joints that traversed the original fresh rock mass.

Human Impacts

Prior to the coastal reclamations that began in the 1860s, the Kowloon Peninsula was long, narrow, and flanked on both sides by a low, sandy coastline, with many muddy inlets. A large sand bar (a spit) was a prominent feature that characterised the southern end of Kowloon. Old survey maps (p. 182) show that the pre-development landscape comprised a series of elongated, low, rounded hills, oriented in an easterly or east-north-easterly direction. The hills were separated by almost straight, gently sloping valleys. Grass and shrubs covered the hills, with trees growing in many of the wide and open stream valleys. Large areas of the open hillsides were scarred by sandy gullies.

Deeply weathered granite underlies the peninsula, giving rise to a distinctive scenery that can still be discerned below the modern urbanisation. Geological mapping (IN52, p. 181) has shown that,

The photograph above shows an area just to the west of the old Kowloon walled city. Numerous burial urns set amid the boulders. In the nineteenth century, hills like these were typical of the landscape. Many were strewn with boulders of varying sizes, which were formed as corestones. These were left on the surface after the intervening, fine-grained weathered materials had been eroded away.

Urbanised Kowloon grew rapidly during the twentieth century, but small hills that once dominated the landscape could still be seen in 2006, as in this photograph from Beacon Hill. The prominent highway is Waterloo Rd. running north-south through the heart of the city. Hong Kong Island can be seen in the background.

IN52 Urban Geological Mapping

Geological mapping consists of identifying the rocks that underlie an area, as well as delineating the distribution of loose surface deposits, such as colluvium on slopes (IN17, p. 91), alluvium in river valleys (IN14, p. 57), and materials placed by people. Making these observations is relatively easy in the natural landscape. However, in urban areas, the shape of the ground has been considerably modified, and the natural features are submerged below a cover of buildings, roads, and other structures. Extensive reclamations around the coast of Kowloon have completely changed the original shape of the peninsula, disguising the underlying geology. Thus, geological mapping in Kowloon requires a different approach.

Cutslopes provide access to geological information that can be used for mapping. In this example the presence of a thick weathered profile and the distribution of corestones can be determined.

Because most of the Kowloon bedrock is concealed, geologists had to focus upon locating particular features. The many cutslopes (artificial slopes) behind the newer housing estates and beside roads were especially useful as these provided exposures of the underlying rock. Unfortunately, in most cases the slopes have been covered with a concrete material called chunam or shotcrete. In contrast, the older urban areas are generally located on lower ground near the former coastline and thus have very few cutslopes. Therefore, geologists also need to look for temporary excavations, such as those dug for skyscrapers. These commonly reach down to bedrock or expose the weathered profile. Similarly, any temporary excavations, such as pipeline trenches, are invaluable.

During the geological mapping of Kowloon (1983–85), MTR tunnels were being constructed and provided continuous exposures of the bedrock. Also, prior to and during World War II, an extensive network of defensive, storage, and communications tunnels were built. These are, in most cases, unlined (the rock has not been covered) and are still in excellent condition. Thus, they provided a great deal of information about the rock types and geological structures.

In many areas, the geologist has to rely on remote methods, the most important source being logs from boreholes drilled to determine foundation levels. There were 15,000 of these at the time of the mapping exercise. Historical maps (from 1845), and aerial photographs (from 1924) of Kowloon, enabled the geologists to examine the pre-development landscape. These sources also helped elucidate the distribution of rock types, geological structures, hillside deposits, river sediments, artificial materials (fill), and the location of the original coastline and its deposits. Sequential aerial photographs allowed the successive reclamations to be mapped and dated. Archival photographs, from the Public Records Office and published in books about old Hong Kong, were also assessed.

prior to urban development, weathered granite was exposed at the surface over about 70% of the peninsula. Over the remaining 30%, the granite was concealed in equal proportion by river sediments and by slope deposits on steeper hillsides. Deep foundations, high cut slopes, and other engineering works can only be designed and carried out (IN53, p. 183) once the thickness and physical characteristics of the weathered materials have been established.

An 1863 map of Kowloon shows some of the first proposed development plans, which included reclamations in the Tsim Sha Tsui bay and along sections of the eastern and western coastlines. Military lands were the first to be demarcated, following which marine and inland lots were sold on short-term leases, beginning in 1864. Reclamation, according to government plans, commenced at the southern tip of Kowloon in 1867, behind a 150 m long seawall paid for by the adjoining lot holders. Similarly, the reclamations at Yau Ma Tei and Tsim Sha Tsui were carried out privately, but to government specifications. Between 1881 and 1883, reclamation was completed, in several lots, to the west of Canton Rd. from Kowloon Point to the then Naval Dockyard (at the end of the present Austin Rd.). Reclamation in Yau Ma Tei

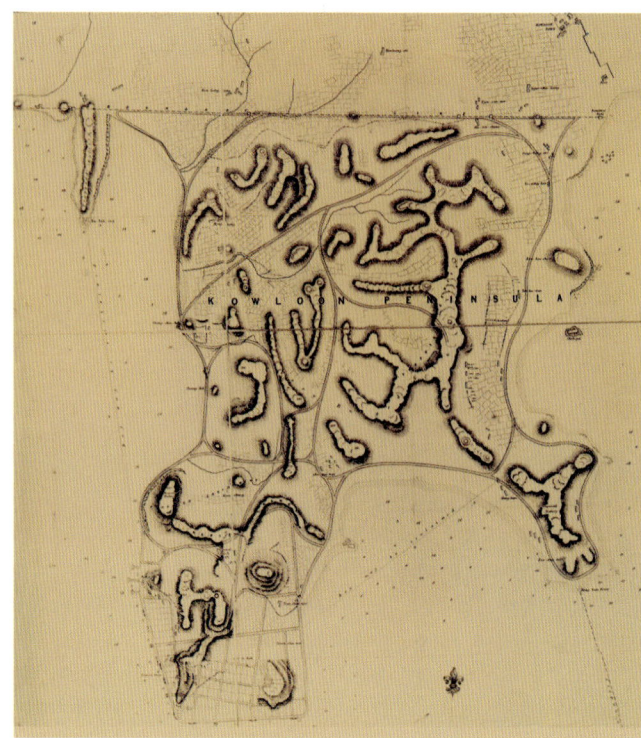

Old maps provide invaluable information about past landscapes. This 1863 map shows Kowloon Peninsula prior to any reclamations, though reclamation plans are noted on the map.

Reclamations along the shores of Kowloon and Hong Kong Island have been incremental. However, increasingly large projects, such as the West Kowloon Reclamation and the shrinkage of Victoria Harbour, have increased public concern that a great natural asset is slowly being lost.

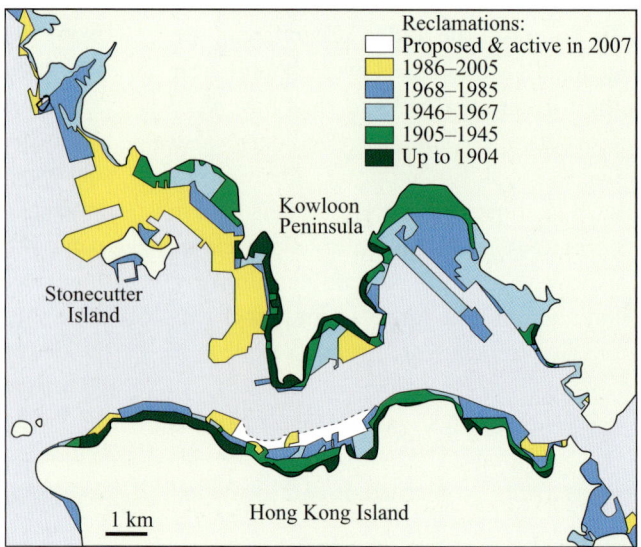

IN53 Urban Geology

Prior to World War II, a minority of the world's population lived in urban areas. Today, about 50% are urban dwellers. United Nations statistics indicate that, of the 326 megacities in the world, the majority are in the Asia-Pacific region.

Sustainable urban development is currently an important objective in many regions, especially in east Asia. Studies have shown that cities of more than one million people are prone to environmental degradation and a variety of natural hazards.

Urban sprawl disguises the underlying natural features, such as flood plains and river terraces, which are still vulnerable to floods. Increasingly scarce land for development means that buildings are being constructed on more marginal sites, such as on steeper slopes and less stable hillsides.

Ground conditions directly influence decisions about the location of buildings and infrastructural facilities, foundation design, water supply, sewage and waste disposal, and flood control. Natural hazards, such as landslides (IN46, p. 159), earthquakes (IN24, p. 108), and tsunamis must also be taken into account. Recently, warnings about global warming have added the magnitude and possible effects of sea level changes to the list of geological factors that need to be considered.

Interestingly, the quarrying of geological materials for producing concrete and building stones is perhaps the most important agent of erosion in the modern world. During 2003, Hong Kong used 16.3 million tonnes of quarried aggregates and other rock products, much of which is imported from mainland China. However, the amount of aggregates and rock products consumed annually in Hong Kong, if quarried within the territory, would be equivalent to removing a layer 6 mm thick from the land surface (an area of 1,036 km^2, excluding reclamations) every year. Clearly, modern urbanisation is having a dramatic impact upon the surface of the earth.

Thus, an increased awareness of urban geology and its role in the safe and sustainable development of modern cities will not only save money but, perhaps more importantly, create a more pleasant urban environment, thereby improving the health and living conditions of the inhabitants.

Hong Kong is a leader in terms of urban geological studies. This has been necessitated by the proximity of high-density buildings close to steep, geotechnically difficult slopes such as those of Fei Ngo Shan (background).

was carried out seawards for 150 m from Reclamation St. Fill was obtained from both the adjacent weathered granite hills and the cuttings made for the developing road network. At Tai Kok Tsui, a series of weathered granite hills were removed, and the material placed in the bay. Other reclamations were completed, privately, at Hung Hom and Ngau Tau Kok.

Reservoirs and Water Supplies

Early European settlers in Hong Kong relied on streams for their water supplies, with the first wells in Kowloon dug at Yau Ma Tei. With the leasing of the New Territories in 1898 and the expansion of Kowloon, the water requirements of the growing city demanded a more secure supply. Consequently, the Public Works Department assigned Mr. L. Gibbs to survey the area now occupied by the Kam Shan Country Park, with the aim of developing a reliable fresh water supply. Work began in 1901 and included a dam, storage tanks, and filter beds. The present Kowloon Reservoir Dam was constructed between 1907 and 1910 and created

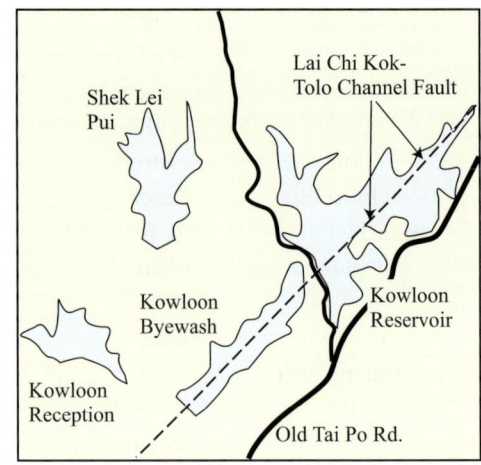

The Kowloon Reservoirs were completed between 1910 and 1931 and were the first major projects in the New Territories to add a new environment to the natural landscape—freshwater, artificial lakes.

The Kowloon group of reservoirs lie within the densely wooded slopes of Kam Shan Country Park. In addition to providing fresh drinking water, the reservoirs are surrounded by numerous walking trails that form a pleasant escape from the nearby city. The dam for the Kowloon Reservoir (upper) was the first to be built in 1910. This was followed by the Shek Lei Pui (middle) and Reception (lower) Reservoirs in 1925 and 1926.

a reservoir with a 1,605 million litres capacity. Further expansion of the system continued with three more dams being completed. These included the Shek Lei Pui Reservoir (527 million litres) in 1923, the Kowloon Reception Reservoir (150 million litres) in 1926, and the Byewash Reservoir (841 million litres) in 1931.

These new water resources introduced fresh aquatic ecosystems to the New Territories, which previously had no large lakes. They also had a profound visual impact on the area, but one that was still controlled by the underlying geology. This contention is clearer when the region is viewed from the air or adjacent mountain tops. The Byewash and Kowloon Reservoirs show a distinctly linear alignment (lower photograph), with the reservoirs occupying a very deep, steep-sided valley. This valley is controlled by the Lai Chi Kok - Tolo Channel Fault Zone (map, p. 184), which separates coarse-grained granite, on the south side of the Byewash Reservoir, from fine-grained granite to the north. The same structure is also of regional importance, determining the orientation of the Sha Tin Valley and the Tolo Channel.

A Mountain Barrier

The Lion Rock Ridge dominates northern Kowloon and has acted as a barrier restricting transport between Sha Tin and Kowloon. Nevertheless, trade over this mountain range has occurred for several hundred years. For example, high-quality incense from the Sha Tin valley was once transported via the Sha Tin Pass to Kowloon, where it was taken on to Aberdeen and, finally, exported to Guangzhou by junk. Charcoal kilns, along the same trade route, once lay on the Sha Tin side of the mountains. Their presence implies, at least, some impact on the forests of the day.

The Byewash Reservoir was opened in 1931, completing the system of reservoirs. Its long, straight form reflects its host valley, which is, in turn, controlled by erosion along a fault line.

Both the Byewash (foreground) and Kowloon (background) Reservoirs lie along a straight line in the landscape. This reflects the controlling influence of a NE-SW fault. Also note that Sha Tin, in the far distance, lies on the same line.

Lion Rock is part of a formidable upland barrier between Sha Tin and Kowloon. The summit rocks are Needle Hill Granite on the north side and Kowloon Granite to the south. Note the near vertical cracks running up the face. These follow relatively easily eroded joint systems within the rock.

During the reign of K'ang Hsi (1662–1773) the coast of southern China was evacuated to a distance of 50 li (29 km) from the sea in order to restrict pirate activity (IN49, p. 165). This clearance damaged the local incense industry, which never fully recovered after re-settlement of the area. During these troubled times, Hong Kong was occupied by troops, and a series of warning fire beacons were established. Included in this defensive system was Beacon Hill. Today, the summit is occupied by modern radar systems for air traffic control.

The Lion Rock Ridge and Kowloon Reservoirs were of vital importance during World War II (IN54, p. 187), when they formed part of the Shing Mun Redoubt, an 18-km long series of defensive positions that was built to defend Kowloon from attack from the north. The line was developed during the mid-1930s, extending from Gin Drinker's Bay (now reclaimed as part of Tsuen Wan) to Port Shelter. The Shing Mun Redoubt, a 48,562-m² site on the north end of Smuggler's Ridge, formed

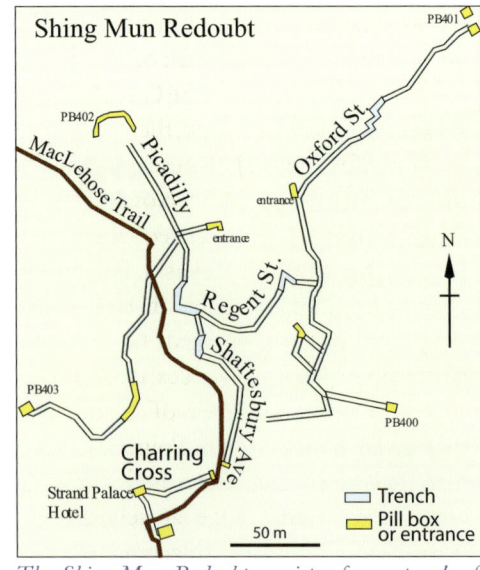

The Shing Mun Redoubt consists of a network of tunnels near the MacLehose Trail. In total they extend for about 18 km and include concrete tunnels and firing points (map, p. 173 for general location).

IN54 World War II

In the mid-1930s, work started on defence positions in Hong Kong. At Devils Peak, for example, old gun emplacements were modernised. The Gin Drinker's Line was developed along natural mountain barriers. This was not a continuous line but rather a series of defensive positions linked by paths and intended to be

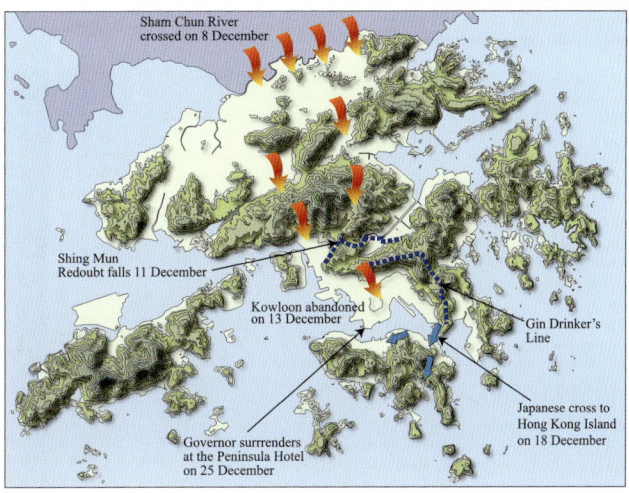

held by 12,000 men; the aim was to slow Japanese forces for at least 3 months, during which time reinforcements might arrive.

Japan invaded China in 1937. Troops occupied Amoy in May 1938 and Guangzhou in October 1938. Hainan Island fell in early 1939. Hong Kong was attacked when the Japanese crossed the Shum Chun River just after 05:00 Hrs. on 8 December 1941. The total army strength in Hong Kong comprised 12,000 men at the outbreak of hostilities, with only 3,500 holding the Gin Drinker's Line. After a night attack, the Shing Mun Redoubt fell and the Japanese troops moved on to occupy Kowloon by 13 December. Subsequently, Lieutenant General Takashi Sakai demanded the surrender of Hong Kong, which was rejected. On 18 December, Japanese forces landed on the northeast corner of Hong Kong Island at North Point and the Tai Koo Docks. From there, the advance met resistance at several forts and gun emplacements. However, most of the gun batteries, such as those at Stanley, served little use as they pointed out to sea, away from the invading forces. Fighting overall lasted for 18 days, after which the Governor signed the surrender documents at the Peninsula Hotel at 15:25 Hrs. on Christmas Day.

Throughout the rest of the war, resistance continued locally. A Chinese group, the East River Guerillas, made use of the mountains to launch attacks.

Vice-Admiral Fujita finally signed the surrender at Government House after the Japanese were defeated in 1945, following the atomic bombing of Hiroshima and Nagasaki.

Old defensive positions are scattered across many of the hills of Hong Kong. This example shows the roof of a pill box along the Hong Kong Trail on Hong Kong Island.

a major base that comprised pill boxes and light artillery positions connected by underground tunnels. The site was abandoned soon after completion due to changes in defence plans but was re-occupied just before the war started.

Today, the Lion Rock Ridge has been breached by road and rail tunnels at several points in order to serve the needs of the growing city. The first link through the mountains, a 2.2-km rail tunnel below Beacon Hill, was constructed for the Kowloon-Canton Railway in 1910. A second rail route was opened in 1981. The first road tunnel was built below Lion Rock and opened in 1967, with a second following in 1978. Other road links were opened in the succeeding years. These included the Shing Mun Tunnel between Sha Tin and Tsuen Wan in 1990, the Tate's Cairn Tunnel between Sha Tin and Kowloon in 1991, and a new tunnel between Sha Tin and Cheung Sha Wan that is due to open by the end of 2008.

The Shing Mun Redoubt tunnels can still be clearly seen today, although they are gradually being reclaimed by trees and scrub. The tunnels, though dark, can be followed for considerable distances.

Country Parks and Wildlife

Three Country Parks protect the northern mountain range. These include the Kam Shan and Lion Rock parks (both designated on 24 June 1978), and the western part of the Ma On Shan Country Park, which was established on 27 April 1979 and expanded on 18 December 1998 (IN28, p. 112).

Long-tailed macaques are a common sight between the Kam Shan and Lion Rock Country Parks. As a result of years of feeding by visitors, they have no fear of people, and can be a nuisance and even bite.

The parks preserve a wide range of wildlife within their well-wooded landscapes. The area around the Eagles Nest Nature Trail (western Lion Rock Country Park), and across the old Tai Po Rd. into Kam Shan Country Park are particularly noted for two species: the long-tailed macaque (*Macaca fascicularis*) and the black-eared kite (*Milvus lineatus*). The latter is a scavenger, relying on offal, refuse, and dead fish from the harbour. The macaques are descendants of monkeys released in the 1920s. They are diurnal and their food includes plants, insects, and small animals.

HONG KONG ISLAND AND LAMMA

Hong Kong Island is the historic heart of Hong Kong. Along with Kowloon, it forms the urban core. The island's landscape displays the widespread imprint of human activities, even in areas that appear natural. Nevertheless, open countryside exists over much of its southern side and on neighbouring Lamma Island.

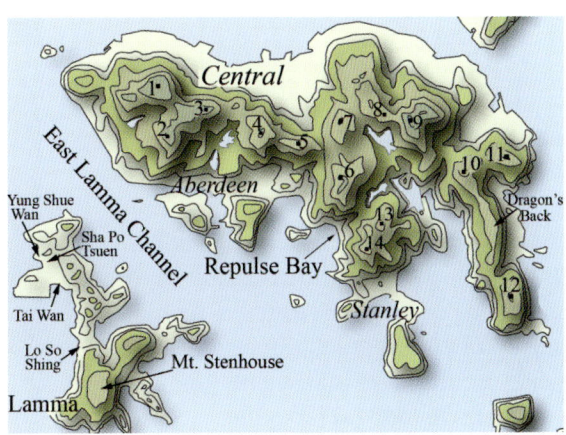

Hong Kong Island was first occupied by British naval forces under the command of Captain Charles Elliot on 20 January 1841. In the following year, the Treaty of Nanking ceded the island to Great Britain in perpetuity. It was returned to China when a 99-year lease on the New Territories expired in 1997. The island is now part of a Special Administrative Region (SAR) of China.

Hong Kong Island (80.4 km^2) is the second largest island in the SAR. In 1841, the population stood at only 15,000, which doubled by 1850. By 2004, the figure had reached about 1.26 million, constituting 18% of Hong Kong's total population. Most of its people and businesses are concentrated along the north shore, between the waterfront and a series of high mountains that include: Victoria Peak (Che Kei Shan), Mt. Kellett, Mt. Gough, Mt. Cameron, Mt. Nicholson, Jardine's Lookout (Ja Din Shan), Mt. Butler (Pat Na Shan), and Mt. Parker (Pak Ka Shan). The hills are steep and valleys exhibit the classic V-shape

Summit	Height (m)	Map No.
Victoria Peak	552	1
Mt. Parker	531	9
Mt. Kellet	501	2
Mt. Gough	479	3
Mt. Cameron	439	4
Mt. Butler	436	8
Jardine's Lookout	433	7
Violet Hill	433	6
Mt. Nicholson	430	5
Stanley Mound	386	14
Mt. Collinson	347	10
Cheung Lin Shan	344	13
Hok Tsui Shan	325	12
Pottinger Peak	312	11

Victoria Harbour is in many ways the pride of Hong Kong. In recent years, there has been an increasing awareness about cumulative damage done to these waters by incremental reclamation. More and more people in Hong Kong have been asking for greater public access to its coastline. This photograph shows the view from the Peak, with Hong Kong's newest skyscraper, Two International Finance Centre, standing high above all others.

created by river erosion in a mountainous area. Streams have been widely modified by the dams and reservoirs built to satisfy the needs of a rapidly growing city.

The vegetation cover reflects the rainfall distribution, wind patterns, and character of the slopes and underlying rocks. Precipitation is higher along the northern mountain spine, reaching more than 2,400 mm per year over the Mt. Butler area. Consequently dense woodlands tend to prevail, especially on the southern mountain slopes where extensive reforestation took place in the later nineteenth and early twentieth centuries. The southern coast of the island is drier, with rainfall being less than 2,000 mm per year. This decreases to less than 1,800 mm in the southeast, where trees are largely absent (except in the lower valleys), and the area is instead characterised by grasses and low shrubs.

The north shore is relatively straight, continuously built up, and contains many of Hong Kong's tallest buildings. In contrast, the south coast is more rural and characterised by numerous pleasant bays

Hong Kong Island is a place of landscape extremes. Remote mountainous countryside (below) lies within close proximity to one of the most exciting urban landscapes in the world (above). The cityscape shows the high density skyscrapers of Central and Wanchai by night from the Peak. The two tallest buildings in this view are the Bank of China building (foreground) and Central Plaza.

Dragon's Back (Lung Check) is a long ridge, about 250–280 m high that extends towards the southeast of Hong Kong Island. It is distinct from much of the rest of the island in being characterised by extensive grasslands and low shrubs. The middle distance summit, at the far end of the ridge, is Mt. Collinson, with Mt. Parker forming the highest peak on the horizon. Pottinger Peak lies to the right and Tai Tam Bay (designated an SSSI) to the left.

Hong Kong Island and Lamma

and long peninsulas. The population densities are much lower, with people living in smaller settlements and towns. The largest urban centres include industrial Aberdeen (with its typhoon shelter and fisher folk), the dormitory town of Repulse Bay, and the tourist centre of Stanley (with its market, restaurants, and a prison).

Lamma (Pok Liu Chau), the third largest island in Hong Kong (13.55 km^2), lies to the southwest across the East Lamma Channel. It has a population of only 6,000, buildings higher than three storeys are prohibited, and there are no roads. Consequently, Lamma provides a much quieter lifestyle for its inhabitants. There are two main settlements: Yung Shue Wan (Banyan Tree Bay) to the north and Sok Kwu Wan (Picnic Bay) on the east coast. The island is hilly and covered mainly with grass or low shrubs. The highest point on the island, Mt. Stenhouse (Shan Tei Tong), rises to 353 m.

The north shore of Hong Kong Island is heavily urbanised and relatively straight. Population densities to the south (right) are lower. The mountainous centre is heavily wooded with mature trees (below) and provides a welcome relief from the harsher city environment.

Lamma lies to the southwest of Hong Kong Island across the deep-water East Lamma Channel, a former major river valley. The island has the shape of two Ys linked in opposite directions. The north Y is Pak A and the south Y is Nam A. The sound of the latter was changed to Lamma.

Landscape Foundations

All of the rocks in this region are igneous, having formed from hot liquid magma. These include several sets of granitic rocks that were intruded at different times and which today underlie about 50% of the area. The oldest granitic rocks are the Tai Po Granodiorite (164 million years old) and the Lantau Granite (161 million years old). Both occur only on southern Hong Kong Island and on southern Lamma. The next to be formed was the South Lamma Granite about 148 million years ago. A final phase of activity produced the D'Aguilar Quartz Monzonite (a granitic rock) and the Kowloon Granite, Po Toi Granite, and Mt. Butler Granite around 140 million years ago (figure, p. 107).

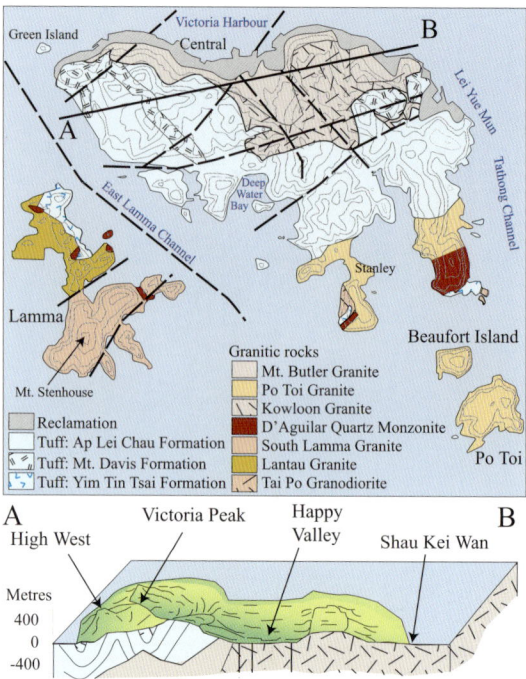

The rocks of the area are of two types: granitic rocks (formed in magma chambers) and tuffs (erupted from volcanoes).

The South Lamma Granite is a medium-grained granite with common black crystals of biotite. It intrudes into the older Lantau Granite that is exposed in the northern half of Lamma. This photograph shows curved slabs of granite and numerous rounded boulders on the upper slopes. Both are highly characteristic of granite landscapes.

Hong Kong Island and Lamma: Landscape Foundations

The remaining rocks are volcanic tuffs. The oldest, which occur on northeast Lamma and the southern parts of the Stanley and Dragon's Back peninsulas, constitute the Yim Tin Tsai Formation (164 million years old). The remaining two tuff formations (named Mt. Davis and Ap Lei Chau) are about 143 million years old.

A detail from the 1:20,000 scale geological map of Hong Kong Island is shown to the right. These maps portray comprehensive information on the rock types. Note how the rocks are aligned in NW-SE-trending bands and how individual rock types are repeated from northeast to southwest. For example, zones of the rock labelled eutaxite appear several times. This is a rock with flattened volcanic fragments that were crushed under the weight of overlying ash while still hot. There are, in fact, only two layers of this rock, but they are folded (IN12, p. 52) into anticlines (upfolds) and synclines (downfolds) as shown in the adjacent section. When the upper rock is eroded away, material in the folds recurs across the landscape.

The Upper and Lower Aberdeen Reservoirs rest on, and are surrounded by, tuffs. These rocks also form the high mountains in this photograph of Hong Kong Island from the south. Mt. Cameron lies at the top centre.

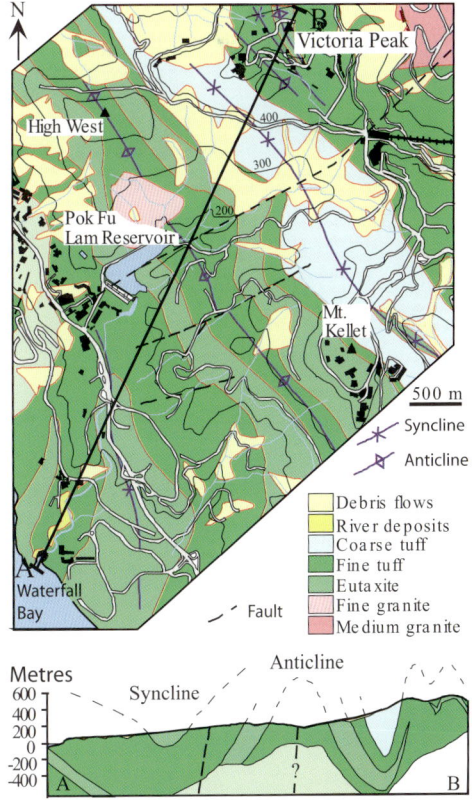

Hong Kong Island from the northwest. The high ground and foreground urban areas are underlain by tuffs, whereas the city to the north (left) of the mountains has been built on granite. Victoria Peak (left summit) and High West (right) mark the position of two anticlines.

These folds are also closely associated with landscape features. For example, the summits of High West and Victoria Peak both coincide with broad anticlinal folds.

Faults weaken the rocks that they cut through, which are then more easily eroded. Consequently, fault lines determine the alignment of valleys. This applies to the Pok Fu Lam Valley and Happy Valley and to the valleys occupied by the Aberdeen and Tai Tam reservoirs. In the case of the latter, the Tai Tam and Tai Tam Tuk reservoirs are located within a NW-SE-trending, fault-controlled valley. In contrast, the Tai Tam Intermediate Reservoir follows a NE-SW fault orientation. These two fault lines intersect at the northern end of Tai Tam Tuk Reservoir. Faults also determine the location of mountain passes (gaps), such as Victoria Gap, Wan Chai Gap, Wong Nai Chung Gap, Mt. Butler Quarry Gap, and Tai Tam Gap.

The Pok Fu Lam Valley (above) follows a NE-SW fault that also controls the location of Victoria Gap in the background. Other gaps and valleys are similarly controlled by faults (below). For example, a NW-SE fault was eroded during times of low sea level and then flooded to form the East Lamma Channel.

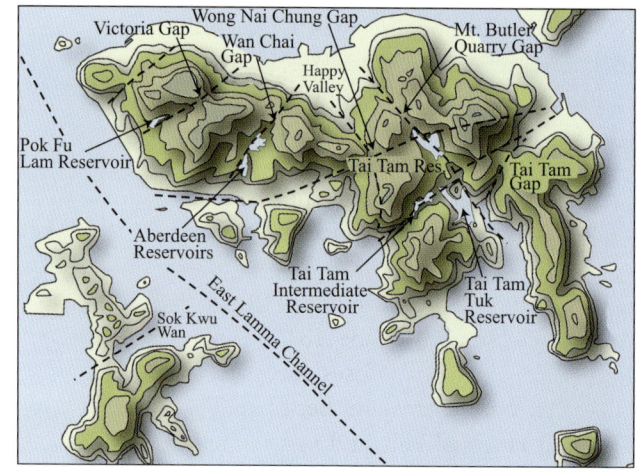

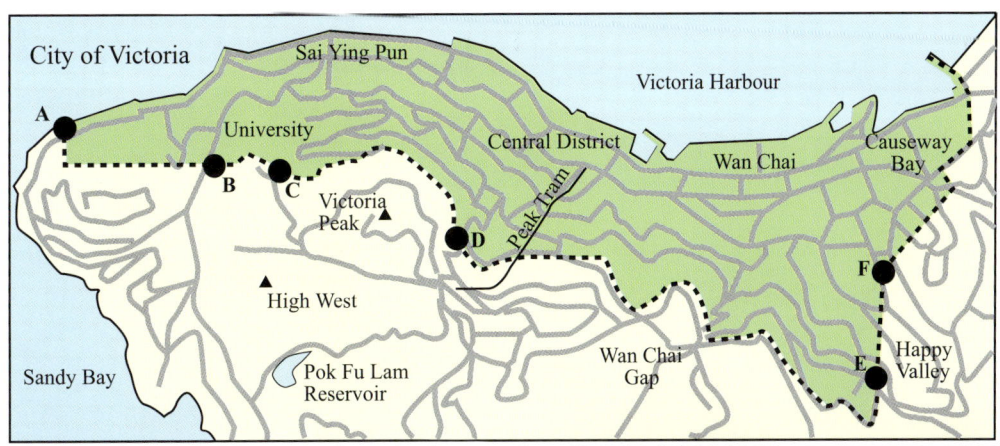

The original City of Victoria is shown in the map above. A series of six granite boundary stones (right) were erected in 1903 to define its limits. Continued urban expansion beyond the original boundary caused them to lose their legal significance. However, all six still exist and, with the exception of the stone in Kennedy Town, are in their original positions. The latter stone was moved to a sitting area in a playground.

Human Impacts

City Origins

Captain Elliot, the Chief Superintendent of Trade, began organising the Colony of Hong Kong in 1841 at a time when there were only 20 villages on Hong Kong Island. Twelve were in the south and eight in the north. The total population was around 5,650 at that time.

The first settlement developed by the British administration was on the northern shores of Hong Kong Island. Originally called Queenstown, the city was renamed Victoria in 1843. Elliot's chief concerns were selecting the main road line (now Queen's Rd.), beginning public works, constructing temporary dwellings for government officials, and issuing preliminary notices of land sales.

The first land sale, comprising 51 land lots, was held on 14 June 1841. Land was sold to 23 merchant houses for offices and godowns. Sale lots were between Sheung Wan and Hospital Hill, and in Sheung Wan,

Stone A: A waterfront playground, Kennedy Town
Stone B: Pokfulam Rd. by lamp post No. 3987
Stone C: Hatton Rd., 400m above Kotewall Rd.
Stone D: Old Peak Rd., 40m above Tregunter Path
Stone E: East Bowen Rd., 500m from Stubbs Rd.
Stone F: Pavement of Wong Nei Chong Rd., opposite St. Paul's Primary School

Central, and Wanchai (map, p. 189). Originally, the proposal was to sell 200 lots, but the number was reduced because of surveying difficulties. By 1843, the Public Works Department had been set up, Crown Leases had been established, and the recommendations of the 1842 Land Committee implemented. Crown land was reserved on the ridge between Glenealy and Albany, which became Government Hill and the site of the Secretariat Office (1848), St. John's Cathedral (1849), Government House (1855), and Albany Government Quarters.

Military land, on which Victoria and Wellington Barracks were built, was set aside between the foot of Government Hill and Wanchai. Murray House Barracks was built in 1844. City Hall and the Hong Kong and Shanghai Bank were constructed between 1860 and 1880. Residential and commercial development

St. John's Cathedral was built in 1849 and given freehold status—the only such land in Hong Kong. The Japanese used it as a Shinto shrine in World War II. It is now a declared monument.

Murray House once stood in Central but was dismantled in 1982 to make way for the Bank of China building. It was restored in 1998 in Stanley and reopened in 1999. Murray House was originally an army barracks, built in 1844, and named after Sir George Murray, the British Secretary of State for War and the Colonies (1828–30). During World War II it was the headquarters for the Japanese military police, with executions taking place in the grounds. Later, it was considered haunted and was exorcised twice.

progressed eastwards along the coast. However, no building was allowed on the military reserve, which resulted in Wanchai and Central being separated.

Civil unrest in southern China in the 1850s resulted in a doubling of population between 1851 and 1855, leading to the opening up of new areas for development and the first land reclamations. By the 1870s, the population had reached 130,168, so further reclamations were undertaken. Increasing congestion prompted Westerners to move their residences up the slope, initially to Caine and Robinson Rds. and later to Conduit Rd. The Peak Tram opened in 1888. The expanding city also generated a need for better medical and university education. Consequently, the Hong Kong College of Medicine was set up in 1877, and this establishment eventually led to the founding of the University of Hong Kong in 1910.

Until 1985, the Legislative Council Building was the Supreme Court. It was built on reclaimed land and opened on 15 January 1912. The blind-folded statue of Justice above the entrance is a replica of the one erected on the Old Bailey in London.

Increasing overcrowding in the city of Victoria prompted the setting up of the Land Commission in 1885. Their chief recommendation was further reclamation. This was carried out in Central between 1890 and 1904, adding 0.25 km² of new land. Population continued to expand, from 262,678 in 1900 to 464,277 in 1911, and to 625,166 in 1920. The 0.4-km² Praya East reclamation scheme in Wanchai was carried out between 1921 and 1931, using material produced by the levelling of Morrison Hill.

The origins of the University of Hong Kong go back to the Hong Kong College of Medicine, which was opened in 1877. Later, the University was formally set up when the foundation stone for the main building was laid by the then Governor, Sir Frederick Lugard, on 16 March 1910.

The earliest major post war scheme was the reclamation of 1951, upon which the present City Hall is located. In 1957, the sale of Murray Barracks, the parade ground, and part of the dockyard released land for development.

Planning began in April 1959, which resulted in the first Statutory Plan for Hong Kong, approved on 11 August 1961.

Over the last few decades, the urbanisation of Hong Kong Island has continued at a rapid pace, with buildings becoming progressively taller. One of the latest contributions to this development, Two International Finance Centre, is 415 m high and was completed in 2003. In early 2007, this 88-storey structure was the highest building in Hong Kong. However, the International Commerce Centre (west Kowloon) was scheduled to rise to a height of 484 m.

Urbanisation, Climate, and Pollution

In general, cities experience higher temperatures (1–4°C) than the neighbouring countryside, an effect called the urban heat island. Also, winds tend to be weaker, shade levels greater, and pollution concentrations higher than in surrounding rural areas.

These patterns are also evident in Hong Kong, where roadside pollution exceeds both regional levels and the relatively lax, governmental, air-quality objectives. In 2007, for example, there was no standard set for the smallest, and most damaging, particulates referred to as PM2.5 (less than 2.5 millionths of a metre).

In addition to these problems, Hong Kong suffers from regional air pollution that frequently reduces ambient visibility. A 2002 government study noted that 80–95% of regional pollutants originated in Guangdong, much of this being derived

From street level, Two International Finance Centre, appears to loom over Hong Kong (above). It is currently the tallest building, has 88 floors, and rises to a height of 415 m.

Queen's Rd. once lay next to the Harbour, but reclamation has since caused the former shoreline to advance northwards. Today, Queen's Rd. is bounded by numerous high-rise buildings and has developed into an "urban canyon", with its own micro-climate and roadside pollution problems.

from Hong Kong-owned factories. This pollution caused poor-visibility days to increase from 15 to 28 days per month between 1998 and 2004.

The regional haze is caused by photochemical reactions, mainly involving sulphur dioxide and nitrogen oxides. These pollutants originate from fossil fuels. Power generation, for example, accounted for about 92% of sulphur dioxide in 2004 (adjacent figure). Of particular concern are factories in Guangdong that generate their own power using low-quality, sulphur-rich fuels. In contrast, road vehicles and shipping are more important as sources of nitrogen oxides (NOx).

Human impacts on Lamma have been severe. The former quarry at Sok Kwu Wan can be seen to the right and in the inset photograph (1997). The Lamma Power Station lies to the left. The quarry provided granite for road and building stone and has now been closed and landscaped. The power plant continues to emit pollutants that contribute to the haze problems.

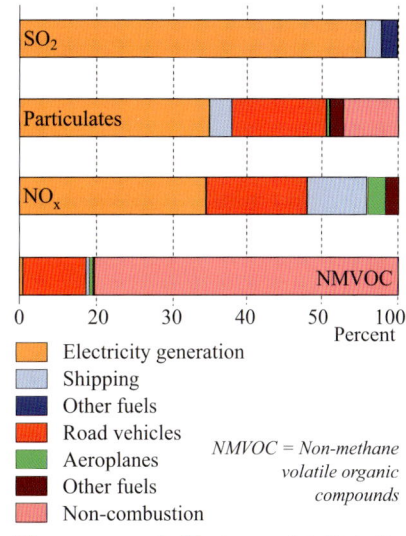

The government's Environmental Protection Department monitors pollutants and publishes data. The information above is for 2004 pollution sources in Hong Kong. The four pollutants pose significant health problems.

Quarries

Quarries have a major visual impact on landscape. Recent discoveries of stone tools, on the shores of Three Fathoms Cove, appear (controversially) to have pushed the record of early human occupation in Hong Kong back to about 35,000 to 39,000 years ago. The abundance of these implements suggest that this site may have been a quarry—the first known in Hong Kong, if the dating is correct. However, at that time, there were few humans, so their impacts would have been minimal.

Urbanisation has led to huge demands for building and road stones, resulting in the opening and closing of many quarries across Hong Kong. Three granite quarries have been active in recent years. The Shek O Quarry, on the D'Aguilar Peninsula, is due to shut down by December 2009. The Anderson Rd. Quarry (east of Kowloon) is visible from the Peak and is being landscaped, with work scheduled for completion by December 2013. The Lam Tei Quarry (north of Tuen Mun) was due to close in March 2007. Two other quarries have been worked in the Hong Kong Island and Lamma region. The Mt. Butler Quarry (1 km east of Jardines Lookout) opened in 1954 and was closed in March 1991. The site is due to be landscaped. Rehabilitation of a granite quarry opposite Sok Kwu Wan (Lamma photograph, p. 199) commenced in 1995 and is now complete. Today, it incorporates woodlands and a lake that have attracted a variety of wildife.

Urbanisation and Streams

In the natural landscape, heavy rain is first soaked up by soils. Once the soil is saturated, surface water flow begins, and stream channels start to fill up. Consequently, there is a delay of several hours between the onset of rain and the streams reaching their maximum flood condition. However, concreting of natural surfaces prevents absorption by the soil. Nearly all of the rainwater then runs directly into streams. As a result, they flood almost immediately and more intensely. Flood controls therefore become necessary because of human interference (middle right photographs). This causes streams to become more erosive and to cut more deeply into their pre-existing channels (adjacent photograph).

The Anderson Rd. Quarry lies to the east of Kowloon but is clearly visible from the Peak. It is currently being landscaped and is due to be fully rehabilitated by the end of 2009.

The two examples above show drainage-control measures on a mountain slope crossed by the Peak Circular Walk. The measure on the left is an open system. The example on the right is an enclosed drain.

Concrete on surfaces above this slope have resulted in a rapid delivery of water to the stream during rainfall and shorter, more intense floods. This increases erosion rates, causing incision of the stream into older deposits.

The dam at the Tai Tam Tuk Reservoir (left) is part of a series constructed between 1889 and 1917. The Pok Fu Lam Reservoir (right) was opened in 1851, but growing demand required expansion to its present size by 1871.

Water supply was a constant problem for the early inhabitants of Hong Kong. Initially, it was derived from streams and, after 1851, from five wells. As the population expanded, streams had to be dammed to supply water. The first Pok Fu Lam Reservoir was opened in 1859, and was replaced by a bigger one in 1871. A large diameter pipe, near the Agriculture Fisheries and Conservation Department (AFCD) Country Park Management Centre, was the main conduit taking water from the Pok Fu Lam Reservoir to Central, giving Conduit Rd. its name.

The Tai Tam Reservoir was started in 1872, but construction stopped in 1874 due to an economic depression. The scheme recommenced in 1882 and was completed by 1889. A 2.5-km tunnel was also built to transfer water to the City of Victoria. Wong Nai Chung Reservoir followed in 1899, with three further Tai Tam Reservoirs being developed between 1904 and 1917. Urban growth led to further expansion of the system and the impounding of the Aberdeen Reservoirs in 1931 and 1932.

In most cases, there was insufficient drainage area within the catchments to fill the reservoirs. Consequently, the water resources were supplemented by constructing a network of gently inclined channels. These catchwaters beheaded many natural water courses on Hong Kong Island, reducing flows in the lower sections of local streams. The Pok Fu Lam

These two catchwater channels intercept streams and divert water to the Upper Aberdeen Reservoir. They also form part of the Hong Kong Trail, a 50-km walk through the countryside of Hong Kong Island.

Reservoir directly intercepted drainage and reduced flows to a famous coastal cascade located at Waterfall Bay (photograph, p. 203). This particular falls was once used by early British sailors to replenish their water supplies. Today, the catchwaters and reservoir paths serve as pleasant, near-horizontal, walking routes through the countryside.

Slope Stability and Maintenance

Slope stability is a major problem in Hong Kong because of the hot humid climate that causes intense weathering and the heavy rains that can trigger landslides (IN46, p. 159). In remote areas, even large landslides cause little or no economic damage, whereas a slope collapse in an urban area can lead to significant loss of life. For example, following six weeks of very wet weather, a major failure occurred on Po Shan Rd. in Mid-Levels at 21.00 Hrs., 18 June 1972. The landslide toppled a 12-storey block of flats and killed 67 people.

Efforts to prevent slope collapse are directed at keeping soil and weathered rock dry. A concrete-like material called shotcrete (adjacent photograph) is sprayed on the surface to keep rain out. Shotcreted slopes usually have protruding pipes (figure) called weepholes that are designed to allow ground water to escape. Drainage channels are constructed above the slope to direct surface water away. In fractured rock, bolts may be drilled into the slope to add further support. Sometimes

This major slump occurred on the north coast of Lamma. Though large, given its location, it caused no loss of life or property.

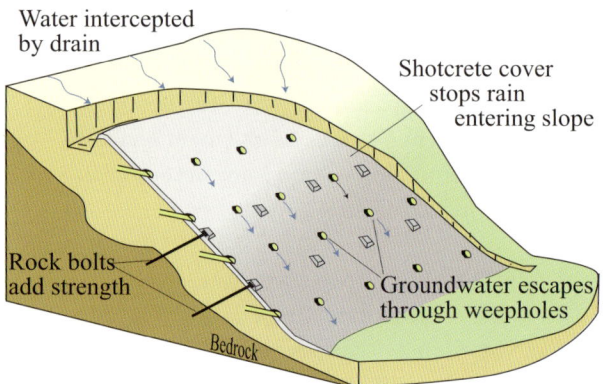

Slope protection methods involve keeping the weathered materials dry by using drainage measures to prevent water from entering the slope. In some cases, bolts are used to anchor the slope materials.

Shotcrete is a concrete-like material that is sprayed onto an incline to prevent rain entering. This slope also shows a large number of protruding, fresh rocks within the loose weathered materials.

This cascade, at Waterfall Bay, was once much larger as it fell over the near-vertical cliff cut into tuffs. The volume of water was reduced after the construction of the Pok Fu Lam Reservoir, which restricted the flow along the lower stream. Note the vertical joints (cracks) that control the form of the well-defined slabs. The falls were first mentioned in the Hsin-an Gazetteer of 1819, which described them as one of the "Eight Scenic Spots in Hsin-an County". They were also an important and reliable watering point for British sailors in the nineteenth century.

Mist nets (above) are used to create a moist microclimate close to the soil or weathered rock. This encourages faster plant growth after a slope is seeded. Stone pitching (right) uses heavy stone blocks to seal and protect the surface.

more substantial stone pitching is used to face a slope (right photograph). In recent years, there have been public objections to the unsightly appearance of this maintenance work, so many inclines are being vegetated, usually with grass. This is encouraged with the use of mist nets that maintain humid conditions suitable for rapid growth after seeding (left photograph).

Archaeological and Historical Landscapes

Archaeological evidence from Lamma suggests that the earliest inhabitants settled there at least 6,000 years ago. These were people, later called the Yueh, who were fishermen that lived near beaches. The date is significant in landscape evolution terms as this is approximately when rising sea levels stabilised at about their present heights. Earlier inhabitants probably existed, but their villages would now lie below the South China Sea.

Lamma was occupied during the Bronze Age, with weapons and tools being found at Sha Po Tsuen, Yung Shue Wan, and Tai Wan (map, p. 189). Tang Dynasty lime kilns survive at Lo So Shing. Both bronze and lime manufacturing require wood, which may indicate the first stages of deforestation. This was evident by the time the British arrived in 1841, when they described Hong Kong Island as a barren rock.

The growth of the city of Victoria from this date had a profound impact on the landscape of Hong

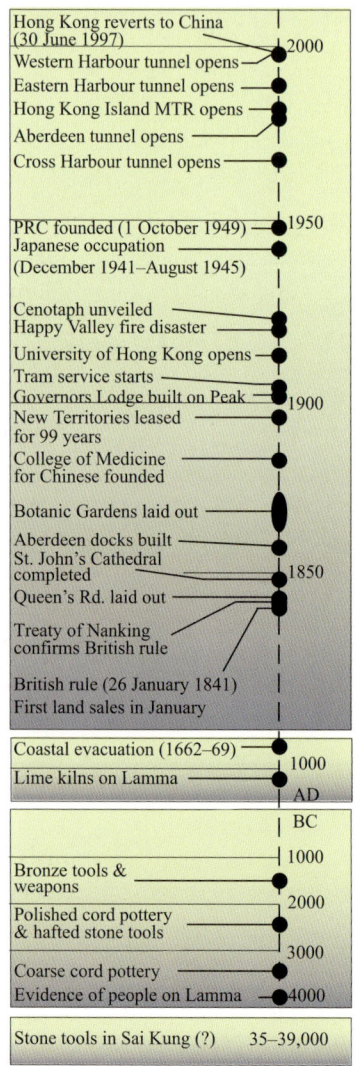

Sham Wan on Lamma is one of several sites where archaeological evidence has been found in or near beaches. The finds here extend back as far as 6,000 years ago. The beach remains important today as a turtle-nesting site that is closed between June and October.

Kong Island. Today, early colonial and Chinese cultural and built landscapes are preserved by the Antiquities and Monuments Ordinance, which came into operation in 1976. Further protection is provided by the Environmental Impact Assessment Ordinance (EIAO) of April 1998. Nevertheless, built landscapes of the past are slowly being lost, for example, the numerous military facilities scattered across Hong Kong Island. In rare cases, such as the British fort at Lei Yue Mun on northeastern Hong Kong Island (built in 1887), installations have been restored. However, in most cases, the facilities are smaller, and many are slowly decaying or being lost under regenerating forests (adjacent photographs).

Wildlife Populations

A wide range of animals are found in the countryside areas, and especially the forested country parks. Pangolins, civet cats, barking deer, Chinese ferret badgers, Chinese porcupines,

World War II defensive positions survive in several parts of Hong Kong's countryside. Many are being lost under a dense vegetation cover.

large bandicoot rats, squirrels, mice and shrews have all been observed. Rhesus macaques were reported in 1963 in Tai Tam, which were possibly remnants of original populations. A 1974–75 survey recorded 44 species of birds. Reptiles are common, though seldom seen (IN55).

The Cape D'Aguilar Marine Reserve (designated on 5 July 1996) lies at the southeast tip of Hong Kong Island. The area is home to fish, corals, and marine invertebrates. Sham Wan, on Lamma, is also ecologically important due to the green turtles that nest there. In order to protect their eggs the beach (adjacent photograph) is closed from June to October.

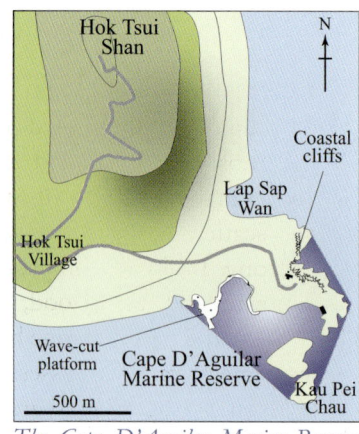

The Cape D'Aguilar Marine Reserve was Hong Kong's first attempt to protect marine environments and is managed by the Agricultural, Fisheries, and Conservation Department, with field assistance from the Swire Institute of Marine Science.

IN55 Amphibians and Reptiles

Amphibians and reptiles, in common with many other animals in Hong Kong, show considerable diversity. Over 100 species have been recorded. The amphibians comprise salamanders and newts, 19 species of frogs, and 3 species of toads. Frogs differ from toads in having slippery skin, webbing between their toes, and generally staying near water. The cascade frog, Hong Kong newt, and Romer's tree frog have been listed under the Wild Animals Protection Ordinance.

Reptiles include lizards, snakes, and chelonians (turtles and terrapins). There are 19 native species of lizards and monitor lizards. In addition, at least 52 kinds of snake, both harmless and venomous, have been recorded. These range from 15 cm to 6 m long. Nine species of chelonians exist, including the green turtle, which breeds locally at Sham Wan on Lamma.

The common toad (top left) and frogs (middle left) are often seen near streams. Snakes such as the banded krait (top right) and the changeable lizard (middle right) are common. The Pacific ridley sea turtle (bottom) died in 2000 at Tai Long Wan Beach.

SEAS AND ISLANDS

This unique region includes some of the most attractive and remote landscapes in the territory—the islands that lie scattered across the seas of Hong Kong. It also encompasses the little-known, or understood, world below the sea. The submarine habitats are extremely varied, possess their own hidden landscapes, and are home to a rich wildlife, albeit one that is threatened by many human activities.

This view from Lantau shows Siu A Chau and Tai A Chau (Soko Islands), in southwestern waters, and is typical of the sea views that add space to the landscape of Hong Kong.

Hong Kong consists mainly of seas (1,800 km²) with land accounting for 1,104 km². Two distinctive traits distinguish the local maritime areas from much of the south China coastline. Firstly, there is an excellent natural harbour north and northwest of Hong Kong Island. Secondly, deep water channels penetrate into the territory, allowing large vessels access to port facilities. These natural factors have encouraged maritime trade, which has been the foundation for Hong Kong's economic growth.

Hong Kong includes 262 islands of varying size. The three largest are Lantau, Hong Kong Island, and Lamma, but there are many more, some of which are named below.

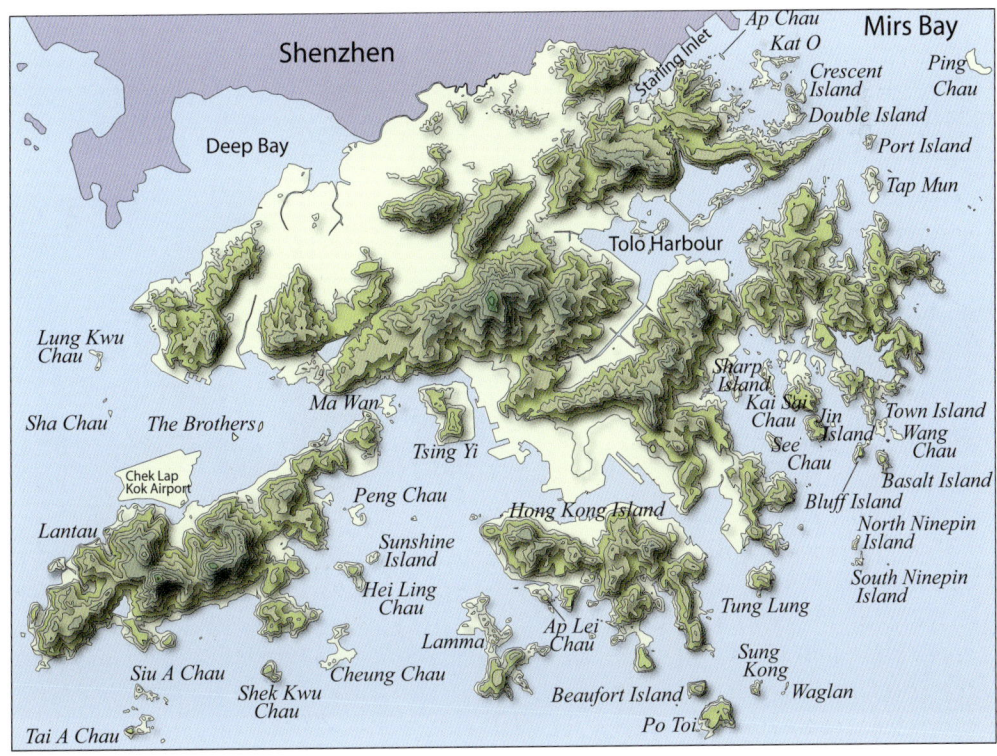

Seas and Islands: Marine Environments

Islands and seas in Hong Kong often provide exhilarating views and isolated beaches. This view was taken from Bluff Island in southeastern Hong Kong. Basalt Island lies to the right and Town Island just to the left.

Marine Environments

In the west, the water is brackish due to the influx of fresh water from the Pearl River, which discharges 350,000 million m³ of water per year. Annually, the river transports more than 80 million tonnes of suspended sediment, about 30 million tonnes of dissolved substances, and an unknown amount of sediment along the sea floor. The extent of estuarine conditions varies seasonally. During the summer, increases in fresh river water and muddy suspended sediment (giving the sea a brownish colour) reflect the monsoon rains. In contrast, drier winter conditions reduce the influence of the Pearl River.

Hong Kong waters change from brackish and muddy in the west to fully oceanic salinities and clear in the east. A transitional region occurs between Lantau and Hong Kong Island. The brackish waters are less dense than the salty marine waters. Consequently, they form a wedge that floats over the top of the denser sea water. Tidal currents, which tend to follow deep-water channels (e.g. the East Lamma Channel), contribute to the muddy water distribution patterns and add complexity to the marine environment.

This view from southwest Lantau shows numerous Chinese islands strewn across the southern Pearl River Estuary.

Seas and Islands: Marine Environments

The variations in salinity and mud content (turbidity) play an important role in the lives of marine organisms. Corals, such as these from Tung ping Chau, are found mainly in the south and east. In contrast, the Chinese white dolphin (IN50, p. 172), for example, tends to be associated with western waters.

The region between Hong Kong Island and Lantau is transitional between the brackish western waters and the fully oceanic conditions of eastern and southeastern Hong Kong. Hong Kong waters are also subject to the seasonally changing influence of several ocean currents (adjacent figure).

These differences are important for offshore ecology. Corals, for example, are confined to eastern and southern Hong Kong, where they exist close to the northern limit of their global distribution. In contrast, the western waters are characterised by mangroves, tidal flats, and notably impressive animals, such as the Chinese white dolphin (*Sousa chinensis*) (IN50, p. 172).

Tides and Tidal Channels

As tides (IN56, p. 210) rise and fall, they generate horizontal movements of water called currents, which play an important role in the marine environment. Current velocities (typically between 0.1 and 2.2 m per second) vary with the shape of the tidal channels. Currents are slower in the wide channels that occur to the west of Lantau and between Lamma and Hong Kong Island. Smaller-scale, deep, narrow channels are present between islands in the east. Here, tidal currents are constricted, increasing their velocity.

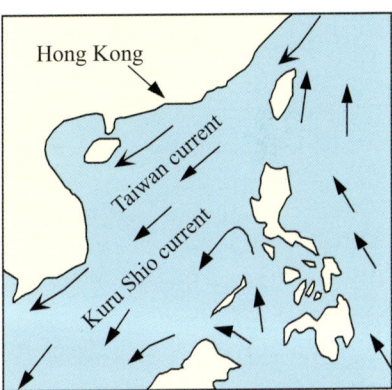
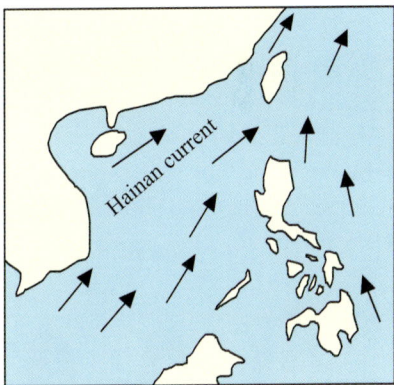

The neighbouring marine areas are dominated by the Hainan current, which flows northwards from the South China Sea with warm surface water and cold, deep, bottom water. In winter, the Kuru Shio current flows from the Pacific via the Luzon Strait, keeping Hong Kong warmer than it would otherwise be. The moderately warm and slightly less salty Taiwan current also flows in winter, but from the northeast, between the mainland coast of China and Taiwan.

IN56 Tides

Tides are caused by vertical changes in sea height due to the gravitational pull of the sun and moon, the influence of the latter being about 2.25 times stronger. The moon, and to a lesser degree the sun, cause two tidal bulges on opposite sides of the earth. During a lunar day, the time taken for the moon to circle the earth (24 hrs. 50 mins.), the sea rises and falls twice, as particular points pass through these tidal bulges. Lunar tides vary somewhat with changes in the angle between the equatorial plane and the moon, and with changes in the position of the moon within its elliptical orbit around the earth. The earth itself revolves once every 24 hours, and the combined effect is to produce a tide once every 12 hours and 12 minutes.

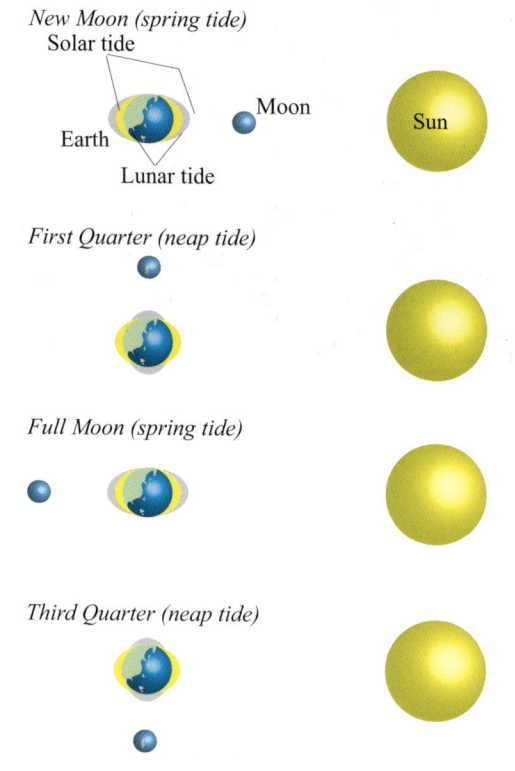

When the earth, sun, and moon are aligned, the combined effect produces very high tides. This happens once every 14.5 days, when high tides are at their highest and low tides are at their lowest. These extreme tides are called spring tides. In contrast, when the earth, sun, and moon are aligned at right angles, high and low tides are less extreme because the two bulges are out of phase. These are termed neap tides.

Land areas also deflect tidal movements, further complicating tidal patterns. In Hong Kong, the tides vary by less than 2 m a day. Two high tides occur per day during spring tides. In contrast, a single high tide develops when neap tides take place. Furthermore, when there are two tides per day they are not of the same height.

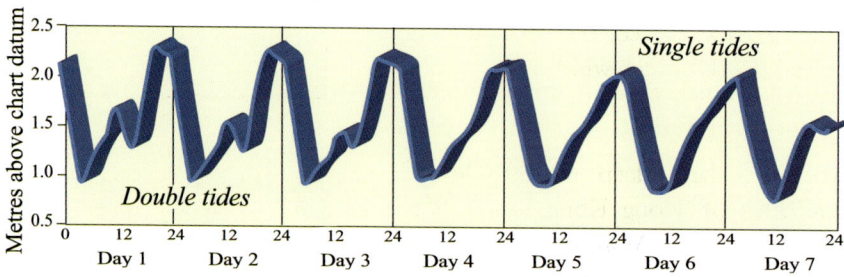

The chart above shows a typical tidal pattern for Hong Kong over a 7-day cycle. To the left there are two high and two low tides of unequal height per day. As time passes, this evolves into a period with just one high and one low tide daily.

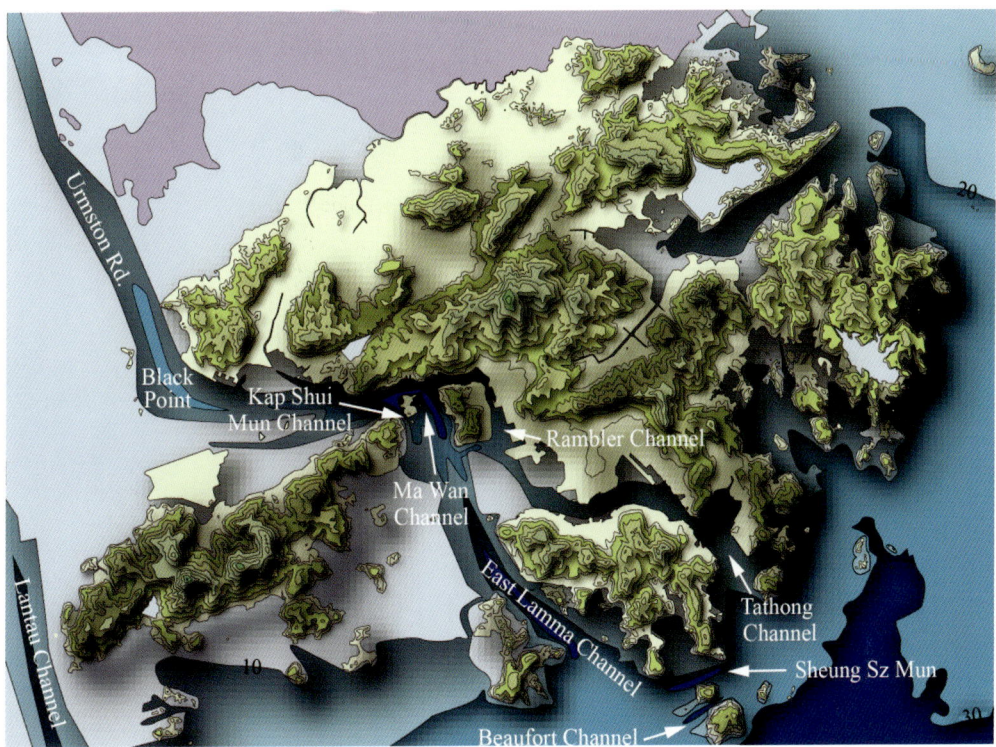

Strong currents pass through the Ma Wan and neighbouring Kap Shui Mun Channels, which open southwards into a wide depression that leads to the West Lamma Channel and the deeper East Lamma Channel. The latter is 40 m deep and a major tidal pathway.

An important tidal channel also passes through Victoria Harbour. This is a southeastward branch from the Rambler Channel, the natural flushing action of which keeps the harbour largely free of sediment. The channel deepens still further to the east, reaching 40 m at Lei Yue Mun, which is rock-floored. It then continues along the Tathong Channel, which is over 20 m deep in places.

Two particularly deep channels occur to the southeast of Hong Kong Island. The sand-floored Sheung Sz Mun, to the north of Beaufort Island, and the rock-floored Lo Chau Mun, or Beaufort Channel, between Po Toi and Beaufort Islands, are both more than 50 m deep. In fact, the Beaufort Channel is the deepest point in Hong Kong waters.

The Urmston Rd. channel crosses the mouth of Deep Bay and is a continuation of the Fanshi Channel of the Pearl River Estuary, reaching depths of over 20 m off Black Point. At the western end of Lantau, the water deepens to over 40 m in the Lantau Channel.

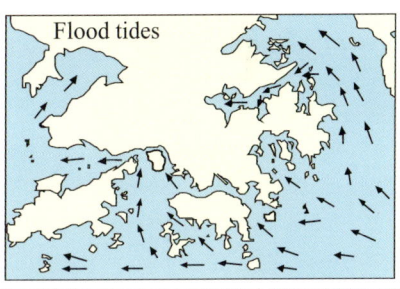
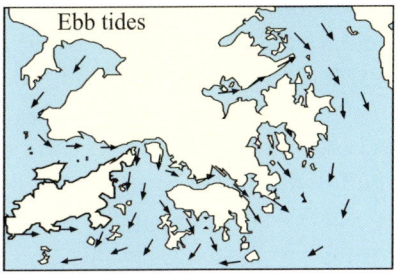

Major tidal pathways are associated with sea floor topography. Note, for example, the flow paths along the East Lamma and Urmston Rd. Channels (compare with large map above).

Seas and Islands: Marine Environments

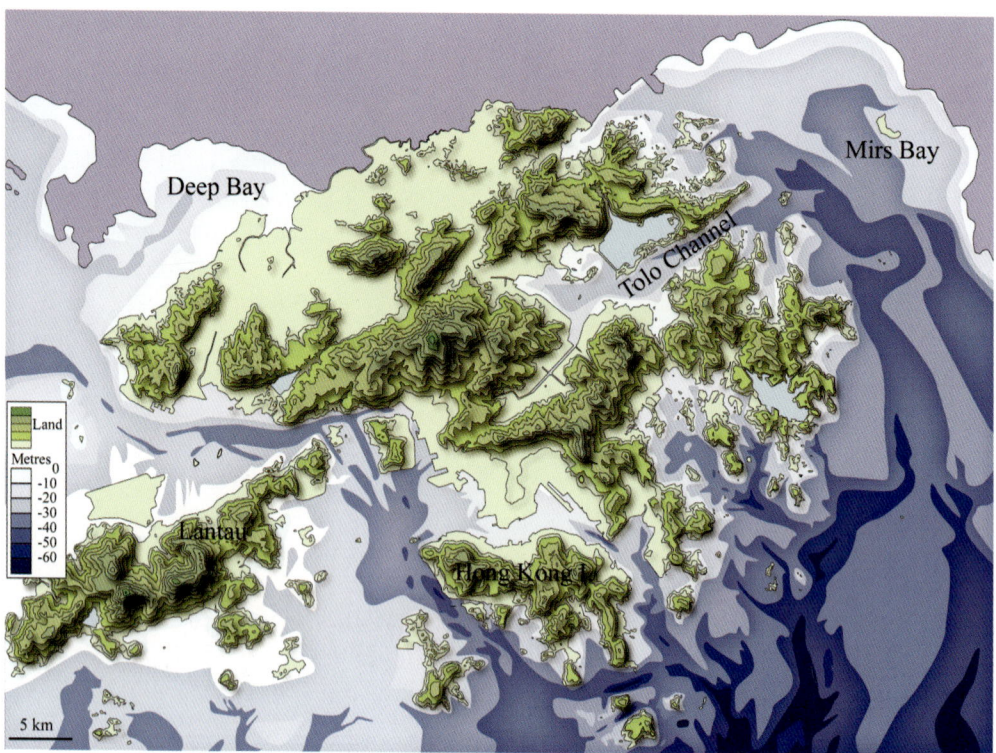

The northern part of Mirs Bay is less than 20 m deep, with water depths increasing southwards. A broad north-south linear depression is incised down to 20 m and extends from the eastern end of the Tolo Channel towards southeastern waters, where the depths reach over 30 m.

Buried Landscapes Offshore

About 25,000–18,000 years ago, the sea level in the South China Sea was 110 m below its present level, and the coastline lay about 120 km to the south of its modern position (adjacent map). Consequently, what is now Hong Kong's near coastal zone was dry land. The rivers that today discharge into the sea at the present shoreline extended their courses southwards by at least 120 km to flow out to sea. These rivers eroded the channels (map above) and deposited sediments across the newly exposed land.

Subsequently, the sea rose, flooding the area, reaching its present height about

Hong Kong waters were entirely dry land 18,000 years ago (map below). At that time, rivers cut deep channels into a broad plain (map above). By 6,000 years ago, the sea had flooded this area, and, at at 5,500 years ago, stood 2 m higher than today. Mud infilled the channels to produce the modern sea floor topography (map, p. 211). A broad bay reached as far as Guangzhou 6,000 years ago (below). Since then, sediments from the Pearl and other rivers have infilled the western side of the bay. This, together with a slight fall in sea level, has produced the modern Pearl River Estuary.

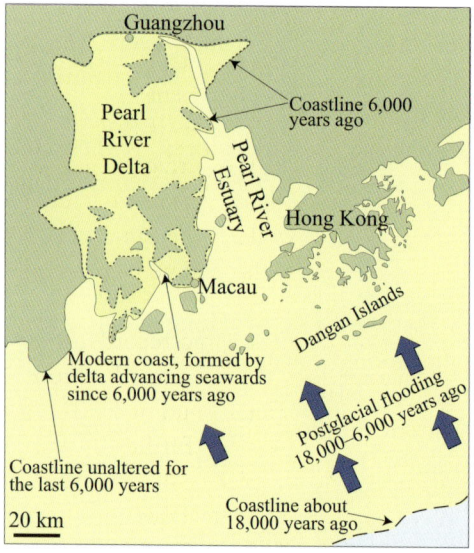

Seas and Islands: Marine Environments

About 18,000 years ago, the Dangan Islands, which lie 20 km south of Po Toi, were hills above a riverine plain that extended 80 km south of the island group.

6,000 years ago. Using seismic profiles (IN57), geologists have reconstructed the old river courses, which are now filled with mud. Some of these channels (e.g. East Lamma Channel) owe their continued existence to strong tidal currents, which have prevented mud being deposited, and thus have left the former waterways exposed on the sea floor.

IN57 Seismic Profiles

Although bedrock can be investigated by drilling boreholes, only the deposits in a thin column below a single point are sampled. Consequently, geophysical techniques are used to build up a clearer three-dimensional picture of subsurface layering.

The most common technique employs a boomer, which is towed behind a boat. It generates sound waves of known characteristics at the sea surface, which then penetrate below the seabed. The signal passes through the first layer of sediment (usually soft mud) but is

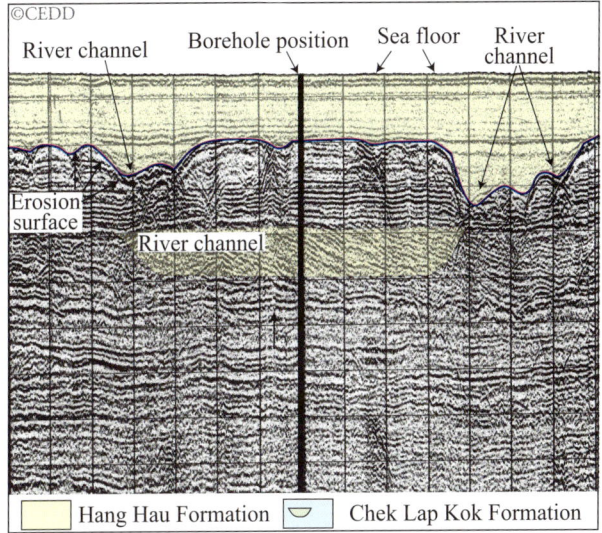

A seismic profile is composed of a series of lines (reflectors) of varying continuity and strength. Note how the Hang Hau Formation, labelled above, is made of faint, continuous, horizontal reflectors—this is marine mud. The inclined layers in the Chek Lap Kok Formation river channel are heavy and discontinous—these are sands. Finally, note how the erosion surface cuts into the layers below—this was formed by downcutting river channels about 18,000 years ago.

reflected back to the surface when it encounters a layer of different density. Hence, the technique is known as seismic reflection. There are further reflections as the signals travel deeper, while at the same time the boat is moving forwards, thus building up a cross-section of the sediment layers.

Importantly, seismic records only show the geometry of the different layers of sediments; they do not provide direct evidence of what the sediments are, although the general character (mud or sand) can be interpreted. Consequently, the second step is to select targets for further investigation by drilling. Careful drilling can retrieve continuously sampled sediments over depths of 80 m or more, enabling geologists to carry out a range of analyses on the sedimentary materials.

Seas and Islands: Human Impacts

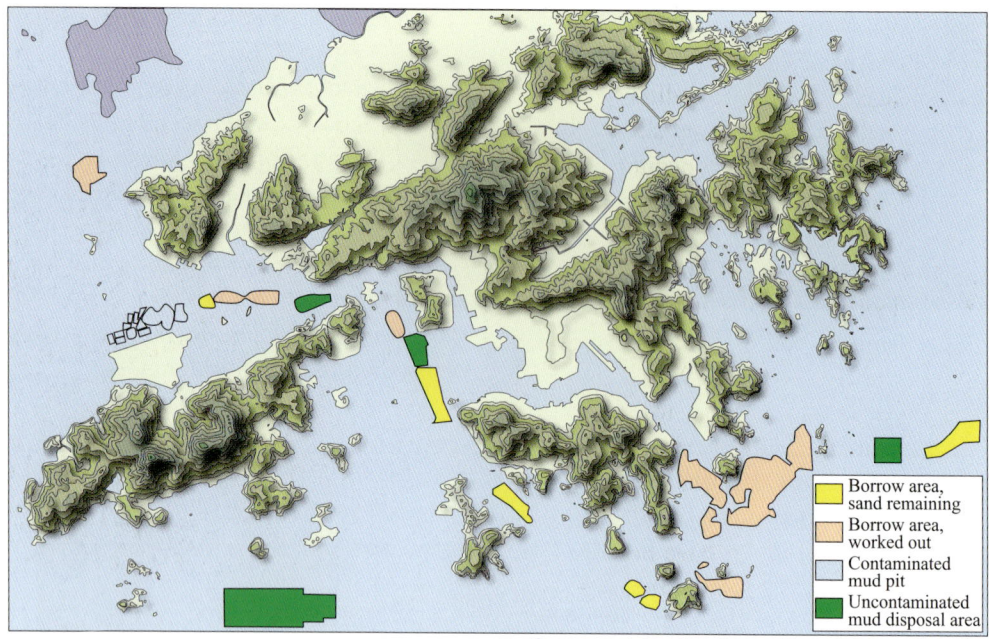

Legend:
- Borrow area, sand remaining
- Borrow area, worked out
- Contaminated mud pit
- Uncontaminated mud disposal area

Human Impacts

Although unseen, the sea floor of Hong Kong has been affected in a number of ways by humans. Several areas of the seabed have been designated as borrow areas, where sediments can be dredged to be used for reclamations or other purposes. Generally, these are locations where sand is exposed. In some cases, the sand has already been removed, whereas at other sites, coarse-grained deposits still remain (map above). Some of these materials (e.g. along the East Lamma Channel and Urmston Rd.; map, p. 211) are old river sands that remain exposed because the channels are conduits for fast tidal currents that have prevented mud deposition. Other sands occur as broad sheets on the sea floor. Some of these accumulations are probably former beach materials that were flooded when the sea rose to its present level.

Other impacts include coastal reclamations, which have converted a total of 67 km² of the sea into land since 1887. Usually, several metres of muddy

Over recent decades, the sea floor of Hong Kong has been modified in many ways by the activities of people. This has occurred through extensive reclamation, trawling (photograph below), dredging for sand, mud dumping or the release of mud into the sea during reclamation operations (bottom photograph), and as a result of cable- and pipe-laying. The map above shows locations that oficially have been designated as sand dredging and mud dumping areas.

overburden first have to be removed in order to build on a firmer (usually sandy) foundation. This dredged fine-grained sediment has then to be disposed of at designated sites. The largest location for dumping of uncontaminated mud is south of the island of Cheung Chau, but several other smaller sites also exist. In some locations (Victoria Harbour, off Tsuen Wan) the sea floor mud was contaminated by dyeing, photochemical and other industrial wastes during the 1960s and 1970s. The factories have now largely moved to the mainland, but the contamination remains. Government regulations require that the polluted mud be disposed of at special sites north of Chek Lap Kok Airport, where pits have been dug on the sea floor to contain the waste materials.

Other impacts are also significant. Nets and weights dragged behind trawlers have left numerous gouge marks on the seabed and, in some cases, damaged seabed ecosystems. Pipe- and cable-laying activities have also affected the sea floor. Marine pollution is widespread and, periodically, gives rise to severe problems. For example, there were 634 red tides between 1980 and 2001. These are caused by excessive growth of algae and other microorganisms (right photographs) when nutrients, such as phosphorous and nitrogen, are washed off the land and into the sea, a process called eutrophication.

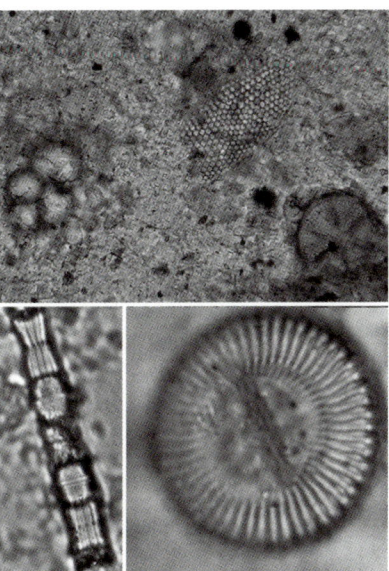

Sea floor mud is a complicated mixture. It consists mainly of clay and silt plus the organic and skeletal remains of microorganisms that floated in the sea while alive. The upper photograph shows mud from Mirs Bay, with a foraminifera on the left. This microscopic animal has a skeleton of calcium carbonate. The two other fossils are diatoms, which are algae with silica skeletons. Diatoms are very sensitive to the environment. For example Cyclotella striata *(lower right) tends to be found in brackish water.* Skeletonema costatum *(lower left) grows in abundance in nutrient-rich water and is one of several organisms that contribute to red tides in Hong Kong.*

The Islands of Hong Kong

The Western Islands

In addition to the Chek Lap Kok Airport platform, there are several small islands in the western waters, north of Lantau. Lung Kwu Chau, Sha Chau, Tree Island, and several small islets lie within the Sha Chau and Lung Kwu Chau Marine Reserve, which was designated on 22 November 1996 to protect the Chinese white dolphin (IN50, p. 172). The islands

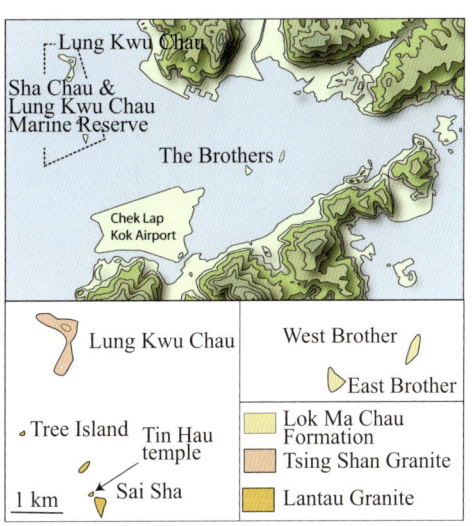

themselves are excluded from this 1,200 hectares park, although they are recognised as Sites of Special Scientific Interest (SSSI), partly because Tree Island (Pak Chau) is an important roosting site for wintering cormorants (*Phalacrocorax carbo*). On Sha Chau, there is a temple, built in 1846, that is dedicated to the goddess of the sea—Tin Hau.

West and East Brother Islands (Tai Mo To and Siu Mo To Chau) lie northeast of Chek Lap Kok. Hills on both islands were levelled to minimise their effects on the airport approach route. The material removed was added to the airport platform. A graphite mine was once located on West Brother Island and formed a link with the early days of the British nuclear power industry (IN58).

Lung Kwu Chau (upper) rises out of the brackish and muddy waters of the Pearl River Estuary. It lies close to the western boundary of Hong Kong and is surrounded by a marine reserve that was set up to help protect the remaining Chinese white dolphins (lower).

IN58 West Brother Island Graphite Mine

Like diamonds, graphite is composed of pure carbon. However, these two minerals differ in the structural arrangement of their atoms. Graphite is a soft, grey mineral and was discovered on West Brother Island by local fishermen, who thought it was coal. Its occurrence was first recorded by Heanley in 1923. The deposits form steeply dipping (60°) seams of graphitic schists (IN11, p. 51), up to 4.5 m thick. The deposit probably originated from plants laid down during the Carboniferous period (354–292 million years ago). Later, these were changed to coal as a result of increasing pressure and temperature following burial. Subsequently, granite intrusions raised the temperatures even higher and altered the coals into graphite-rich schists.

The Tin Bo Mining Development Co. started working the deposits in late 1952. Later, the mine was run by the Ng Fuk Black Lead Mining Co. Ltd. The graphite was initially sent to the UK and later to the USA, probably for manufacturing control rods used in nuclear power stations. By the late 1950s, the mine was producing up to 3,500 tonnes of graphite annually. By the 1960s, the mine had reached depths of 90 m, and continuous pumping was required to keep it dry. The mine closed in January 1973.

Graphite schist (above) is a shiny grey rock. Originally a coal, it was changed by hot magma. The map below shows the steeply dipping graphite seams. Note how the hard quartzite (a sandstone) controls the northern ridge. The island has now been flattened as part of the airport project.

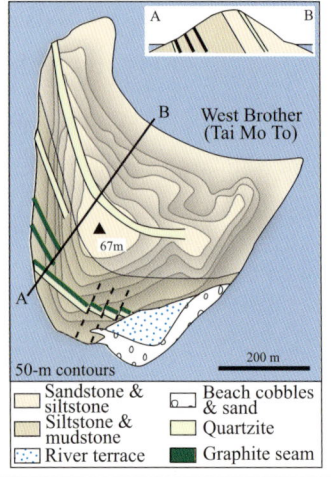

Seas and Islands: The Islands of Hong Kong

The Soko Islands are in the foreground of this view from Lantau. The first larger island is Siu A Chau, with Tai A Chau just behind. The third large island, and others in the distance, lie within mainland waters. The geology of the group is comparatively simple, with most islands being made of granite or rhyodacite (maps below).

The Soko Islands

The Soko Islands are located in the southwest of Hong Kong, in an area subject to the brackish water influence of the Pearl River Estuary. The islands are composed mainly of granite and, consequently, have a generally rounded form. A belt of rhyodacite dykes (p. 154–155) that passes through Tai A Chau is associated with low ground, upon which a Vietnamese Detention Centre (IN59) was erected. This closed down in September 1996.

IN59 Vietnamese Refugees

Following the 1975 fall of Saigon, many thousands of Vietnamese escaped in hundreds of small boats. Those who landed in British Hong Kong were sent to detention camps across the territory. The largest was at Whitehead (near Ma On Shan), with other smaller facilities placed in relatively remote locations, such as the Soko Islands. Altogether, about 2.5 million people left Vietnam, Laos, and Cambodia for neighbouring countries such as Indonesia, Malaysia, the Philippines, Singapore, and Thailand. Many moved on to the USA, Australia, or Europe. The population of the Hong Kong camps dropped from 18,886 in April 1996 to 1,801 in August 1997. The government closed the Tai A Chau Detention Centre in 1996 and the largest camp, at Whitehead, in June 1997. A small number of Vietnamese remained in Hong Kong.

Siu A Chau viewed from the east, with Sum Wan at the lower left. Note the rounded hills, which are typical of a granitic landscape.

Seas and Islands: The Islands of Hong Kong

Proposals have been made to designate the waters around the Soko Islands as a marine reserve. They are ecologically rich, with coral communities (right) and marine wildlife, such as starfish and rays (above).

The seas around the Sokos are ecologically rich and are home to Chinese white dolphins and finless porpoises, as well as fish, corals, rays, starfish, crabs, and other species. Proposals have been made that the area should be designated a marine reserve or an ecotourism destination. A less environmentally sensitive suggestion is for a new island to be created, 3 km west of the Sokos, as a way of disposing of waste materials; it would be landscaped at a later stage. Another environmentally questionable proposal in 2007 was that a Liquid Natural Gas terminal be built on Tai A Chau. The future of this island group remains uncertain.

The Islands off Eastern Lantau

Geologically, Peng Chau, Kau Yi Chau, Sunshine Island, and Hei Ling Chau consist mainly of granite, which was originally formed in magma chambers 161 million years ago. Rhyodacite dykes, with large feldspar crystals (IN45, p. 155), were intruded into the granites about 15 million years later. Cheung Chau and Shek Kwu Chau are composed of the Chi Ma Wan Granite, a younger suite of rocks that were intruded 143 million years ago. Cheung Chau actually consists of two islets linked by sand, a feature referred to as a tombolo.

Peng Chau (literally Flat Island) is a small isle (0.98 km²) that is home to about 8,000 people, with fishing as a major industry. It is famed for its seafood and religious sites, hosting a Tin Hau temple dating back to 1792. There are excellent views from Finger Hill (95 m), the highest point (top photograph, p. 219).

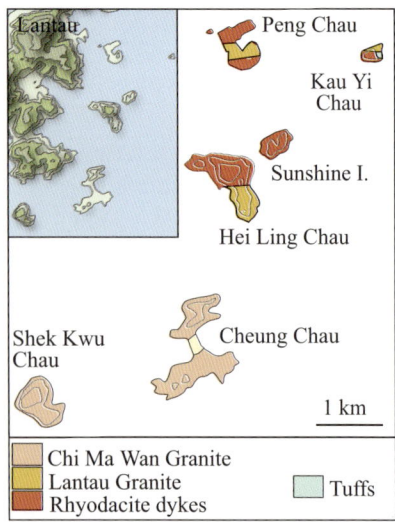

Islands off the east coast of Lantau are dominated by two types of granite. The northern islands are also cut by numerous, parallel dykes of rhyodacite.

Peng Chau (top) consists of two granite areas separated by dykes, which underlie the village in this view from Finger Hill (south Peng Chau). Hei Ling Chau is the larger island in the centre left photograph. Dykes form Sunshine Island to the right and the northern (near side) of Hei Ling Chau, the remainder of which is granite. Shek Kwu Chau (centre right) and Cheung Chau (bottom) are made of younger granites, with a sand tombolo underlying the village on the latter island.

Seas and Islands: The Islands of Hong Kong

Only two of the six islands lying to the east of Lantau are permanently inhabited by active communities. Fishing harbours, narrow streets, local shops, and seafood restaurants are characteristic features of both Cheung Chau (above) and Peng Chau (right).

Cheung Chau (Long Island) has an area of 2.45 km², a long history of occupation, and a thriving village. The population stands at about 30,000. The central part of the island is developed with shops, narrow lanes, and a recently expanded harbour that also acts as a typhoon shelter.

Hei Ling Chau (1.93 km²) is home to a drug addiction treatment centre, set up in 1975. In 2004, the government proposed that a "super jail" be built on the island together with a bridge to Lantau, an environmental nightmare that would have cost HK$12 billion. The idea received sustained opposition, and was abandoned. Neighbouring Sunshine Island is a breeding habitat for white-bellied sea eagles.

White-bellied sea eagles are sometimes seen flying overhead in this maritime setting. Breeding pairs are thought to occur on Sunshine Island, having left Lantau's Penny's Bay due to development of the Disneyland resort.

The Southern Islands of the Po Toi Group

The Po Toi Granite was one of the last large magma bodies to be intruded, cool down, and solidify. This occurred about 140 million years ago, and erosion has since brought these rocks to the surface. They now form nearly all of Beaufort Island, Po Toi (3.69 km²), and Sung Kong. The older Lantau Granite (161 million years old) constitutes the rocks forming Waglan in the east of the island group. Several dykes (IN45, p. 155) occur on Po Toi and

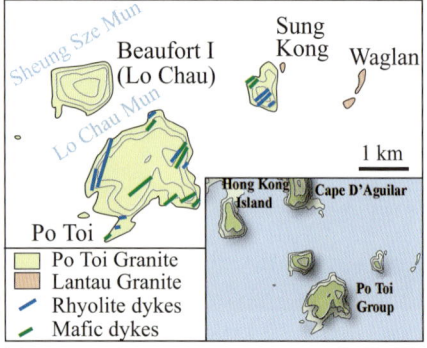

Two types of granite are present in the Po Toi group of islands, which are also affected by dyke intrusions on Po Toi and Sung Kong.

Beaufort Island (left) is separated from lower and broader Po Toi by the narrow Lo Chau Mun (right), which is the deepest point in Hong Kong, with depths exceeding 60 m.

Sung Kong (adjacent photograph). These are of two compositions. Firstly, fine-grained, dark grey, basaltic dykes that formed between 108-75 million years ago. Secondly, quartzphyric rhyolite dykes (IN45, p. 55) of unknown age. These are also fine-grained, but are light coloured, and contain quartz crystals. They follow the same northeasterly trend as the basaltic dykes.

Rock exposures are common on all of the islands, with rounded boulders littering the landscape. The hills have rounded summits and convex profiles, especially on the upper slopes (photograph below). Large sheeting joints (IN09, p. 37) occur parallel to the smooth hill sides, forming curved slabs that sometimes detach and slide down slope.

This thin, fine-grained, dark grey, basaltic dyke cuts through Po Toi Granite, which contains large crystals of feldspar and quartz, and minor amounts of biotite (IN21, p. 100).

Very well-developed sheeting joints (IN09, p. 37) occur in granite outcrops along the southern coast of Po Toi. These joints have dictated the convex form of the slopes in this area.

Waglan consists of older granites and has an irregular form, with erosion focused on vertical sets of joints.

The view north from the summit of Po Toi reveals a classic granitic landscape of rounded hills, scattered outcrops, and isolated boulders. Lamma lies to the left, with Hong Kong Island under the clouded backdrop to the right.

Human impacts on the Po Toi Group are minimal. Once a home to about a thousand people, only about 200 live on Po Toi today. Rock carvings along Nam Tam Wan on the south coast have been declared monuments and consist of lines and spirals said to resemble stylized animals and fish. Beaufort and Sung Kong are uninhabited. Waglan is host to Hong Kong's oldest lighthouse (1893). It has been unoccupied since 1989 and was recognised as a monument on 29 December 2000. Today, it also acts as a weather station.

The Southeastern Islands

Tuffs dominate all of the southeastern islands, but they erupted at different times and in somewhat contrasting settings. On the Ninepins, the tuffs belong to the High Island Formation, which was formed about 141 million years ago. These particular islands were designated a Site of Special Scientific Interest (SSSI) on 16 February 1979 because of the excellent columnar joints (IN40, p. 136) that occur there. These were developed within the Sai Kung Caldera (IN40), a large circular depression formed by the collapse of a volcano. Hot volcanic ash that accumulated within this structure

The Tin Hau temple at Tai Wan, Po Toi.

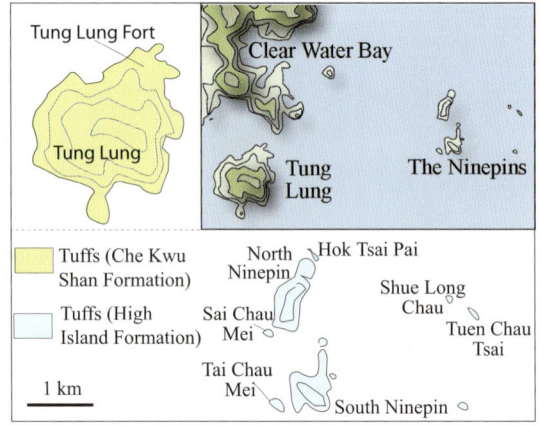

Tuffs occur on all of the islands, but those on Tung Lung are older. World-class columnar joints characterise the Ninepins.

222

Seas and Islands: The Islands of Hong Kong

This photograph shows the rugged and steep North Ninepin Island, with its three distinct hills. The small islet of Hok Tsai Pai is in the foreground. These isles lie in waters exposed to powerful and erosive ocean waves.

cooled slowly, contracting to form large hexagonal columns that today shape the ground surface.

The rocks on Tung Lung Island (Che Kwu Shan Formation) are slightly older (about 142 million years) and formed as a result of very violent eruptions. Exposure of these rocks to wave action has had a marked effect on the island's landscape. Most areas are characterised by smooth, gentle-to-steep, grassy slopes. In contrast,

North Ninepin from South Ninepin. Sai Kung Country Park and the Basalt Island group are on the horizon.

Well-developed columnar jointing is widespread through the Ninepins. These joints on South Ninepin are mostly five- or six-sided. Note how the slope of the surface follows the form of the columnar joints.

the southeastern shores are exposed to ocean waves and have been eroded to produce cliffs that plunge into the adjacent sea. Joints within the rocks have been eroded to form a jagged and irregular scenery. Undercutting by waves has triggered failures, some of which are very large.

All of the southeastern islands are uninhabited today. However, Tung Lung Island (also known as Nam Tong) was once occupied by Chinese soldiers who built a fort on its northern side, about 300 years ago. The fort was set up, along with others in the region, in order to protect trade and to defend against marauding pirates (IN49, p. 165). Country park status has been suggested for the island, which is home to an amazingly diverse flora. Greater protection might also be needed for the world class columnar joints and marine habitats of the Ninepins. Recently, proposals have been made that a wind farm be built in shallow water close to these islands.

The scenery on Tung Lung Island varies with exposure to ocean waves. Most of its slopes are smooth and grassy (above), though they can be steep. In contrast, the southeast coast consists of rugged sea cliffs and slot caves that have been eroded back along joint lines (below).

The Islands of Rocky Harbour and Port Shelter

The inner islands, between the waters of Port Shelter and Rocky Harbour, have a gentler aspect than those in the more exposed southern areas. All of these isles consist of tuffs, mostly belonging to the High Island Formation (adjacent map), which is 140.9 million years old. Very slightly younger tuffs of the Clear Water Bay Formation (140.7 million years old) prevail on the more westerly Sharp Island and Shelter Island.

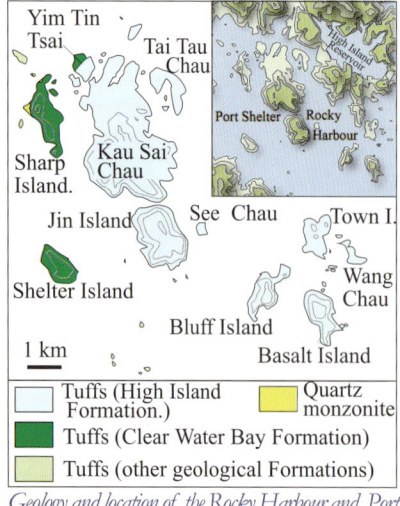

The numerous inter-island channels make for pleasant boat excursions. Formerly, these were river channels, and the islands were the intervening hill tops. The channels were flooded as glaciers melted and returned water from land to the sea.

The rugged outer isles include: Town Island (Fo Tau Fan Chau), Wang Chau, Basalt Island, and Bluff Island (Sha Tong Hau Shan). The

Geology and location of the Rocky Harbour and Port Shelter groups of islands.

Town Island is broad and low, rising to only 72 m. The indented coast and attractive cliffs are cut by caves at many points, most of which are associated with joints or faults. Note the arrows that mark an east-facing shady cliff on the island and on Pak Fu Shan in the background. Although they seem to be offset (a persepective effect), they actually lie along the same line when viewed from above. This is a fault that has controlled their development.

latter two isles possess some of the most dramatic scenic locations in Hong Kong. Bluff Island, for example, is particularly noted for its 140-m-high sea cliffs, the tallest in the territory. Both Bluff and Basalt Islands were designated as Sites of Special Scientific Interest in February 1979 because of their very well-developed columnar jointing (IN40, p. 136).

The undersea landscapes are equally stunning, with corals

North- and south-facing coastlines vary considerably in the outer islands. The upper right photograph shows the highly eroded southern coast of Bluff Island. Contrast this with the north-facing aspect, which is protected from ocean waves (lower).

Seas and Islands: The Islands of Hong Kong

The coast of Basalt Island is a popular diving spot. Below the waters in the photograph is a wealth of marine life, with anemones (A and B, below right) being abundant. This and other parts of Basalt Island are also home to the Pentagon Cucumber (C), other types of sea cucumbers (D), sea apples (E), Moorish idol (F), and puffer fish (G). Bluff Island has a similar variety that includes sea cucumbers (H) and anemones (I). Small squid (J) and sea urchins (K) are present around Sharp Island, and octopus (L) can be found off the nearby eastern coast of Sai Kung.

occurring in many areas. Parts of the coastline around Bluff Island are half covered by corals. Fields of anemones sway in the currents below Basalt Island (photograph above). An enormous range of marine life lives in these varied habitats (adjacent photographs).

The inner islands include: Kau Sai Chau (the largest at 6.69 km^2), Jin, Sharp, and Shelter islands, as well as Tai Tau Chau, See Chau, and Yim Tin Tsai, plus many smaller islets. Human impacts on these islands are more significant than on the outer isles. The northern part of Kau Sai Chau, for example, has two 18-hole public golf courses, the first of which opened in 1995. Construction has started on a third course. A small fishing village lies at its southern extremity, next to Jin Island.

Villagers first settled on Yim Tin Tsai about 300 years ago, with the local population peaking at about 500, although few remain. Catholic missionaries converted the local population and built St. Joseph's chapel in 1878–9. The church has since been restored and is the oldest example of its kind in Hong Kong. A local salt industry once existed, with most of the produce being sold in Sai Kung. Salt production ceased long ago. Today, Yim Tin Tsai, Tai Tau Chau, and the inter-island gap between Kau Sai Chau and Jin Island serve as fish farming (mariculture) zones.

Sharp Island (Kiu Tsui Chau) is mostly designated as a Country Park—the smallest in Hong Kong. The island is thickly wooded and lies within the waters of Inner Port Shelter (Sai Kung Hoi).

Kiu Tsui Country Park lies on Sharp Island and is the smallest (1 km²) in Hong Kong. The island differs from others in the area because it possesses a variety of rock types. These include quartz monzonite (a quartz-poor granite, p. 86), volcanic sediments, and banded lavas, as well as the more common tuffs.

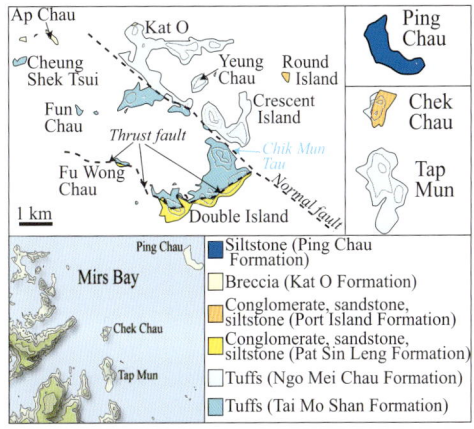

The Islands of Mirs Bay

Mirs Bay is blessed with some of the most varied islands in Hong Kong; Chek Chau, Ping Chau, and Kat O were declared Sites of Special Scientific Interest (SSSIs) in 1979. This largely reflects the geological variety of the rocks of which they are composed (map), which include tuffs (as elsewhere), but also sandstones, siltstones, conglomerates, and sedimentary rocks composed of angular blocks (breccias).

The neighbouring islands of Tap Mun (Grass Island) and Chek Chau (Port Island) demonstrate these contrasts well (adjacent photographs). Tap Mun is composed of tuffs and has been eroded to a broad rounded shape because there is little variation in the rocks to produce any preferred slope directions. In contrast, Chek Chau is composed of several types of sediments (sandstones, siltstones, and conglomerates) of the late Cretaceous (99–65 million years ago) Port Island Formation (IN06, p. 29). These rocks dip (slope) to the east and have controlled the shape of

The upper photograph shows Tap Mun from Chek Chau. The isle is broad and rounded, contrasting with the asymmetrical form of Chek Chau (foreground and lower photograph), caused by its dipping sedimentary rock layers.

the island, with gentler slopes to the east and steeper scarps to the west. At smaller scales, harder rock layers stand out on the hillsides, whereas softer rocks are eroded to produce depressions (IN07, p. 31).

On the opposite side of Mirs Bay, to the northeast, lies Ping Chau, the most isolated island in the territory. This isle has a particularly distinctive form, with a gentle slope to the east and a steep cliff and wave-cut platform (p. 74) on the west. The island is made up of thinly layered siltstones of the Ping Chau Formation (IN06, p. 29), which were originally laid down in an ephemeral desert lake during the early Tertiary, about 60 million years ago.

Ap Chau lies at the northern end of Crooked Harbour and comprises large angular blocks (up to 10 cm) of breccia. These form gently dipping layers that slope northwards, controlling the shape of the island. The layering, and its relationship to the slopes, can readily be seen in a sea arch at the north end of the island (adjacent photograph). The rocks form part of the Kat O Formation (IN06, p. 29) and originally accumulated in an arid setting, where the angular blocks fell off a north-facing fault scarp. This cliff was located a short distance to the south of the island, but cannot be seen today as it has long since been eroded away.

The remaining isles are mostly composed of tuffs of the Tai Mo Shan Formation (164 million years old) or Ngo Mei Chau Formation (142 million years old). These rocks

The westerly facing cliffs of Ping Chau cut across finely-layered siltstones. In this photograph the layers can be seen in the cliff and on the wave cut platform below. Note how the island slopes gently away to the east (right). This surface simply follows the gently dipping rock layers.

Ap Chau is about 500 m long and is very narrow, with gently dipping layers of sedimentary breccia. These can be seen sloping from right to left in the lower photograph of a sea arch. Note how both the hill and the arch broadly follow the rock layers in form. The arch is located at the bottom end of the island in the upper photograph and is surrounded by a supratidal platform (p. 77).

Seas and Islands: The Islands of Hong Kong

are separated by a vertically oriented normal fault (IN08, p. 35), which has been eroded to form the Chik Mun Tau channel (map, p. 227). In contrast, a gently dipping thrust fault (IN08) passes through Double Island. This slopes downwards to the north, and has forced older tuffs over the top of younger sandstones (cross section). The trace of the thrust fault is shown by the dashes in the right-hand photograph. The rocks also dip gently to the north. Consequently, the island has an overall asymmetrical shape, with steeper slopes along its southern coast.

Today, most of the islands are uninhabited. In the past, Double Haven (photograph above) was a centre for the collection of pearls, which were transported to Tuen Mun for onward shipping to Canton. The waters of this bay are now partly protected by a marine park (p. 81). In other areas, local activities include mariculture (practised at Tap Mun, O Pui Tong, and Kat O Wan) and trawling.

The only remaining sizeable communities are on Tap Mun and Kat O. Small villages are present on Ap Chau and Ping Chau. Tap Mun once had 5,000 people, but there are now less than 100. The last school closed in 2004. Diving is popular, and travellers stop at Governor's Beach, on Double Island, for swimming and relaxation.

Double Island (Wong Wan Chau) is the largest in this group, which also includes Kat O and Crescent Island. Ap Chau lies just beyond the left edge of this photograph.

Sedimentary rocks form the southern coast of Double Island. Tuffs occur above these, having been moved on top of the sediments (yellow in the cross section) by a major, low-angle, thrust fault (IN08, p. 35). This is shown by the dotted line in the photograph above and in the cross section, which extends from Ap Chau in the north, through Double Haven, to Double Island in the south.

Double Island possesses a fine bay and beach (Governor's Beach), fed by sediment from a stream draining the central hills of the island. A brackish water lagoon is blocked by the beach.

Epilogue:
Landscapes, Past, Present, and Future

Hong Kong is a small territory situated on the south coast of China. Yet it contains a remarkable range of natural and human-made habitats. The region is mountainous, with numerous intricate valleys. Lowlands occur in several regions. The vegetation is highly varied, as is the fauna. The imprint of people is widespread. Examples include tea terraces on hillsides, rice paddies on lowlands, and extensive urban areas, many on reclamations.

In order to understand the origins of this scenic diversity, it is important to realise that landscapes are not just the result of geology or formed by a single process, or developed only by human activities. Rather, they reflect, to varying degrees, the combined action of all three influences.

Geology is the fundamental basis for understanding the story of landscape in Hong Kong. However, it is also important to consider the role of time in landscape evolution. The scenery we see today has developed over millions of years. Events that occurred long ago control the form of the modern landscape, which is carved out of ancient materials created in the distant past. The preceding chapters have shown how violent volcanic activity shook the region at about 180 million years ago and again between 164 and 140 million years ago. During these periods, volcanic ash was spewed out in great quantities, sometimes from gargantuan eruptions, as rocks were swallowed along a major boundary in the earth's crust. At these times, there were rapid changes in landscapes as volcanoes arose and were then eroded away. However, landscape evolution, over most of Hong

An understanding of the origins of landscapes involves many factors. For example, this small valley near the northeast coast of Plover Cove Country Park is seldom visited, but shows several landscape types. Streams have carved the valley out of tuffs formed many millions of years ago. The valley, along with many others in Hong Kong, has been partially flooded by rising sea levels that allowed mangroves to grow. Dense secondary forests have replaced the original tree cover that was lost following settlement of the area over the last few hundred years.

This photograph shows part of the Tai Lam Reservoir. Although this appears to be a simple landscape, multiple influences need to be considered in understanding the scenery. The rocks are decayed granites, the result of natural weathering over millenia. Erosion followed deforestation by humans, which exposed the weak rocks to stream incision and gullying (natural processes). The forests are secondary and were deliberately established to reduce erosion and prevent sediment filling the reservoir—a water supply for the growing population.

Kong's 400-million-year geological record, has been slow and steady, but unrelenting.

In Hong Kong, the rocks are dominated by tuffs, granites, and a variety of sedimentary lithologies. These vary in their resistance to decay and erosion, so that some stand high, while others form low ground. Faults weaken rocks and thus determine the orientation of valleys. The dominance of northeast-southwest valleys in Hong Kong reflects this important control on landscape. Some rock layers are horizontal, others slope gently, and a few stand nearly vertical. The orientation of these layers commonly determines the shape of the ground surface, bringing about both symmetrical and asymetrical landforms. Where there is no layering, hills tend to be more rounded in form.

The landscapes that exist today have been carved out of the rocks by surface processes that operate today, as well as the geologically recent past. Ultimately, these processes are driven by energy from the sun. For example, water evaporates from the sea and eventually falls on the land. Exposure to this water and to the atmosphere causes rocks to decay chemically. The loosened materials can then move downhill, either slowly, or more rapidly as a landslide. Rain water contributes to rivers, which may erode away the surface materials or deposit sediment. Along the coast, erosion and deposition are caused by wave action, which, in turn, results from winds. Changes in climate and sea level similarly reflect variations in solar energy received by the earth.

Chemical decay of rocks, as well as river, marine, and slope processes, has operated for a long time, so the landscape we see today reflects their cumulative effects. These surface processes can also change in their magnitude with time. For example, during the past two million years, and, in particular over the past 35,000 years, glaciers built up in the circumpolar regions. As a result, major sea level fluctuations occurred that caused a temporary imbalance in the processes of erosion and deposition that were operating over the landscape. Several times, as the surface of the oceans fell, rivers began to cut down to a new lower sea level, only to have to re-establish their former valley floor levels, by deposition, as the sea rose during periods when the glaciers melted back. Since the beginning of the Holocene, about 10,000 years ago, natural earth surface processes have been directed towards redressing this imbalance, attempting to create a new quasi-equilibrium. The modern landscape is thus partly inherited from the past and reflects the varying influence of many factors.

Then people arrived on the scene. For many centuries, Hong Kong was little affected by people. Suddenly, several waves of settlers arrived and began to make alterations to the countryside. Hong Kong had entered a new phase of landscape upheaval that began almost imperceptibly at first but which gradually gathered momentum as the population swelled and as technology advanced. Even from the start, the rate of change far outweighed most natural processes that had occurred over previous millenia. Agricultural terraces on the hillsides; widespread forest clearance; the building of walls; houses, and trackways; damming of rivers; and draining of marshes—all had a marked impact upon the scenery. Thus, people began to move geological materials, modify drainage patterns, change vegetation cover, and hunt wildlife.

The development of more advanced technologies saw the magnitude of these

Epilogue

Urbanisation exerts a direct human impact on landscapes. The photograph shows rapidly developing Tung Chung. This New Town did not exist prior to the opening of the new airport (to the left of the photograph) in 1997.

modifications increase exponentially. Heavy machines, with the ability to move ever-increasing quantities of rock and soil, expanded the scale of the alterations. Thus, from the early, small, public reclamations of the late nineteenth century, which were carried out over several years, human-made change along the coasts culminated in the massive reclamations of West Kowloon and Chek Lap Kok at the end of the twentieth century; these were completed in only a matter of months. The landscape of Hong Kong had been modified permanently. Urbanisation has led to the development of dramatic city landscapes that have partly been built on these extensive reclamations but which also extend across natural lowlands, and climb up lower mountain slopes.

The rocks and faults that form the ground are not going to change rapidly in the future. River, marine, and slope processes are unlikely to vary dramatically. Apart from the unlikely occurrence of a major earthquake or tsunami, future landscape change (on a human time scale) is likely to reflect both indirect (global warming) and direct (urbanisation) human impacts on landscape.

Global Warming

Global warming raises global average temperatures. The main environmental effects of this are, or would be, to raise sea levels and to increase the frequency of extreme weather events. In addition, rainfall patterns are likely to change.

Sea level has already risen more than 120 m since the last ice advance about 18,000 years ago. Most of the rise occurred before 6,000 years ago. From about 3,000 years ago to the beginning of the nineteenth century, sea level rose about 0.1 to 0.2 mm a year. Since 1900, it has been estimated that the sea has risen by about 1–2 mm a year. However, modern methods of satellite measurement indicate a rise now of about 3 mm a year.

Hong Kong has a densely developed coastal fringe that would be prone to flooding if the sea should rise. Only a minor increase would make the effects of high spring tides and storm surges a greater threat. Reclamations would be particularly vulnerable. A significant increase in the height of the sea would radically change the shoreline in the northwest, where there are very low coastal lowlands.

Typhoons would become more common as ocean temperatures increased. If combined with a rising sea level, they would pose a greater threat to low-lying regions. Related natural effects would be higher storm-surge flooding, leading to increased coastal erosion and changes in shoreline habitats.

Temperature changes are implied by global warming, but rainfall is also likely to alter. However, it is difficult to predict if conditions will be drier or wetter. In either case, there probably would be modifications in the vegetation that covers the landscape. River and floodplain processes would also change in response to any increase or decrease in the amount of water available.

Urbanisation and Landscape

By the end of 2004, the population had reached 6,895,500. In the same year, 38,800 immigrants entered the territory and there were 21.8 million visitor arrivals. Thus, during 2004, three times the resident population entered Hong Kong. Such large numbers place enormous pressures on the natural resources.

The increasing population needs to be housed. During 2004, 47,000 new residential units were constructed. Except in redevelopment areas, these projects demanded land. Between 1995 and 2004, there were 57 infrastructure projects to support housing. Continued urbanisation has major implications for air and water quality, noise levels, and waste disposal.

During 2004, the community disposed of 9,290 tonnes of municipal solid wastes every day. On average, each person threw away 1.36 kg of waste per day. Currently, there are three landfills in Hong Kong, with a predicted life of only 6–10 years. These are due to be landscaped once they are exhausted. Pressures for replacement sites are increasing, so searches are continuing for new areas that will be changed radically.

Large populations also create pollution problems. Sewage, for example, was disposed of through untreated outfalls into Victoria Harbour. This system has now largely been replaced with a new underground sewage network comprising 23.6 kilometres of tunnels. Rapid urban development is currently producing construction and demolition debris amounting to 56,000 tonnes each day. Of this, 88% is suitable for re-use in reclamation projects. The

Waste is a major landscape problem. Unofficial dumps, as shown above are unsightly, but the major problem is the total volume of waste. This creates a demand for landfill sites that is not sustainable.

remainder has to be disposed of elsewhere, again taking up valuable land (or seabed). Development projects also generate vast quantities of dredged marine mud that is unsuitable for reclamation or other uses. A total of 31.3 million m^3 was produced in 2004, which was dumped at sea in specially designated dumping grounds.

In addition to land being swallowed up for housing, commercial premises, and industry, land is also occupied by transport networks. In 2004, there were 1,943 km of roads. These are continuing to expand. For example, a bridge linking Hong Kong-Zuhai-Macao is currently being investigated.

In recent years, there has been increasing public pressure to scale-back the planned major new reclamations, several of which appear to have road improvements as their primary objective.

Finally, the problems of a polluted atmosphere cannot be ignored in landscape terms. The sources and types of pollutants are varied. Some originate within Hong Kong, and others come from mainland China, often from industries that are run by Hong Kong owners. These pollutants cause health problems, but they also obscure the scenery. Low visibility is defined locally as meaning that it is possible to see clearly no more than 8 km, and excludes periods of rain, fog, and mist. In 2004, low visibility occurred about 18% of the time. More frequently, there is a general haze, especially in winter, that degrades landscape views. A landscape that cannot be seen is no longer an asset to the people of Hong Kong.

Hong Kong Island viewed from Lion Rock, with Kowloon in the foreground. The photograph was taken on a relatively good day for visibility. Commonly, Hong Kong Island cannot be seen. Even on this good day, note the brown pollution haze over Hong Kong Island and the Western Anchorage (to the right).

INFORMATION SOURCES AND FURTHER READING

This work has made use of an extensive range of data sources, as well as the authors' own previous works. Major sources are listed in Part I, with other references that have been used less intensively in Part II. Part III includes minor source materials and other books and articles about Hong Kong, grouped into categories for ease of reference. The selection is not comprehensive but will allow readers to pursue individual topics of interest.

PART I: Publications Referred to Extensively

Hong Kong Animals by D. Hill & K. Phillips, Hong Kong Government Printer, 1981, 281p.

Hong Kong Country Parks by S.L. Thrower, Hong Kong Government Information Services, 1984, 216p.

Hong Kong Landscapes, Along the MacLehose Trail by R.B. Owen & R. Shaw, Geotrails, 2001, 203p.

The Pre-Quaternary Geology of Hong Kong by R.J. Sewell, S.D.G. Campbell, C.J.N. Fletcher, K.W. Lai, P.A. Kirk, Hong Kong Geological Survey Memoir, 2000, 181p.

The Quaternary Geology of Hong Kong by J.A. Fyfe, R. Shaw, S.D.G. Campbell, K.W. Lai, P.A. Kirk, Hong Kong Geological Survey Memoir, 2000, 209p.

The Geology and Exploitation of the Ma On Shan Magnetite Deposit by P.J. Strange & N.W. Woods, Geological Society of Hong Kong Newsletter, 1991, 9(1), pp. 3–15.

The Geology and Exploitation of the Needle Hill Wolframite Deposit by K.J. Roberts & P.J. Strange, Geological Society of Hong Kong Newsletter, 1991, 9(3), pp. 29–40.

The Geology and Exploitation of the West Brother Island Graphite Deposit by N.W. Woods & R. L. Langford, Geological Society of Hong Kong Newsletter, 1991, 9(2), pp. 24–35.

The Story of Lin Ma Hang Lead Mine, 1915–1962 by T. Williams, Geological Society of Hong Kong Newsletter, 1991, 9(3), pp. 3–27.

PART II: Publications referred to less extensively

1. Geological

Distant-earthquake Simulations Considering Source Rupture Propagation: Refining the Seismic Hazard of Hong Kong by K. Megawati & A.M. Chandler, Earthquake Engineering and Structural Dynamics, 2006; 35, pp. 613–635.

Geology of North Lantau Island and Ma Wan by R.J. Sewell & J.W.C. James, Hong Kong Geological Survey Sheet Report No. 4, Geotechnical Engineering Office, Hong Kong Government Printer, 1995, 46p.

Geomorphological Observations on Rainwash Forms in Hong Kong and some other Humid Regions of Southeast Asia by H-L Tschang, The Chung Chi Journal, 1972, 11, pp. 40–59.

Geotechnical Area Studies Programme—Territory of Hong Kong by K.A. Styles & A. Hansen, Geotechnical Control Office, Civil Engineering Services Department, Hong Kong. GASP Report, 1989. No. XII, 346p plus 15 maps.

Hong Kong Geological Survey Memoir No. 1: The Geology of Sha Tin by R. Addison & R.J. Purser, Hong Kong Government Printer, 1986, 85p.

Information Sources and Further Reading

Stone Trail Waterfall, Sai Kung East Country Park

Hong Kong Geological Survey Memoir No. 2: The Geology of Hong Kong Island and Kowloon by P.J. Strange & R. Shaw, Hong Kong Government Printer, 1986, 134p.

Hong Kong Geological Survey Memoir No. 3: The Geology of the Western New Territories by R.L. Langford, K. W. Lai, R.S. Arthurton & R. Shaw, Hong Kong Government Printer, 1989, 140p.

Hong Kong Geological Survey Memoir No. 4: The Geology of Sai Kung and Clearwater Bay by P.J. Strange, R. Shaw & R. Addison, Hong Kong Government Printer, 1990, 111p.

Hong Kong Geological Survey Memoir No. 5: The Geology of the Northeastern New Territories by K.W. Lai, S.D.G. Campbell & R. Shaw, Hong Kong Government Printer, 1996, 144p.

Hong Kong Geological Survey Memoir No. 6: The Geology of Lantau District by R.L. Langford, J.W.C. James, R. Shaw, S.D.G. Campbell, P.A. Kirk & R.J. Sewell, Hong Kong Government Printer, 1995, 173p.

Preliminary Vegetation Maps of the World since the Last Glacial Maximum: An Aid to Archaeological Understanding by J. M. Adams & H. Faure, Journal of Archaeological Science, 1997, 24, pp. 623–647.

The Influence of the Pearl River on the Offshore Geology of the Macao—Hong Kong area by R. Shaw & J.A. Fyfe, Proceedings of the International Conference on the Pearl River Estuary in the Surrounding Area of Macao, 1992, pp. 247–255.

The Map That Changed the World: A Tale of Rocks, Ruin and Redemption by Simon Winchester, Penguin Books, 2002, 238p.

Updating of Hong Kong Geological Survey 1:20,000 Maps. Major Findings and Revisions Map Sheet 7—Sha Tin by R.J. Sewell & J.C.F. Wong. GEO Report No. 179, 2006, 29p.

Urban Geological Mapping—Techniques used in Kowloon and Hong Kong by P.J. Strange, In: The Role of Geology in Urban Development, Geological Society of Hong Kong Bulletin No. 3, 1987, pp. 181–189.

Urban Geology and the Impact on Our Lives: Samples from Daily Life in Bangkok. Atlas of Urban Geology: Volume 13 by United Nations Economic and Social Commission for Asia and the Pacific, United Nations, New York, 2001, 80p.

Variations in Sub-tropical Weathering Profiles over the Kowloon Granite, Hong Kong by R. Shaw, Journal of the Geological Society, London, 1997, 154, pp. 1077–1085.

2. Natural History

A Pilot Biodiversity Study of the eastern Frontier Closed Area and North East New Territories, Hong Kong, June-December 2003 by Kadoorie Farm and Botanic Garden, Hong Kong Special Administrative Region, 2004, 67p.

Birds of Hong Kong by C. Viney & K. Philips, Hong Kong Government Printer, 1986, 214p.

Conservation of Corals in Hong Kong by P.O. Ang, Proceedings of IUCN/WCPA-EA-4 Taipei Conference March 18–23, 2002, Taipei, Taiwan, pp. 277–295

Hong Kong Amphibians and Reptiles by S.J. Karsen, M.W.N. Lau & A. Bogadek, Hong Kong Urban Council Publication, Hong Kong Government Printer, 1986, 136p.

Hong Kong Insects by D.S. Hill & W.W.K. Cheung, Hong Kong Urban Council Publication, Hong Kong Government Printer, 1988, 128p.

Hong Kong Insects: Volume II by D.S. Hill, Hong Kong Urban Council Publication, Hong Kong Government Printer, 1982, 144p.

Hong Kong Island and Po Toi Island, Friends of the Earth, Coastal Guide Series, 1998, 29p.

Hong Kong Mangroves by N. Tam & Y. Wong, City University of Hong Kong Press, 2000, 148p.

Hong Kong Trees: Omnibus Volume by S.L. Thrower, Hong Kong Urban Council Publication, Hong Kong Government Printer, 1988, 438p.
Lantau Island, Friends of the Earth, Coastal Guide Series, 1997, 27p.
Lamma Island, Friends of the Earth, Coastal Guide Series, 1995, 19p.
The Corals of Hong Kong by P.J.B. Scott, Hong Kong University Press, 1984, 112p.

3. Human Impacts and History

A History of Hong Kong by F. Welsh, Harper Collins, 1993, 624p.
An Illustrated History of Hong Kong by N. Cameron, Oxford University Press, 1991, 362p.
Chinese Graves and Gravemarkers in Hong Kong by C. Chow & E. Teather, Annual Journal of the Association for Grave-stone Studies, 1998, pp. 286-336.
City of Victoria: A Selection of the Museum's Historical Photographs by The Urban Council of Hong Kong. Hong Kong Museum of History, 1994, 107p.
Feng Shui by S. Rossbach, Rider, London, 1990, 169p.
Forts and Pirates: A History of Hong Kong by Y.C.A. Lui, K.K. Siu, T. Stanley & C.N.C. Lui, Hong Kong History Society, 1990, 114p.
Forts and Batteries: Coastal Defence in Guangdong During the Ming and Qing Dynasties by K.K. Sui, Urban Council of Hong Kong, 1997, 125p.
Hong Kong's New Towns: A Selective Review by R. Bristow, Oxford University Press, 1989, 385p.
Hong Kong—The Colony That Never Was by A. Brich, Odyssey, 1991, 160p.
Martyrs, Mystery and Memory Behind the Colonial Shift by S.C.H. Cheung, ASA Conference, 2000, School of Oriental & African Studies, London Participating in Development: Approaches to Indigenous Knowledge 2nd–5th April 2000, 19p.

South Lamma Coastline

Political Disintegration of Hakka Villages: A Study of Drastic Social Change in the New Territories of Hong Kong by M.I. Berkowitz & E.K.K. Poon, Chung Chi Journal, 1969, 8(2), pp. 16–31.

Remains of War by J. Wordie, South China Morning Post Magazine, 21 November 1999, pp. 10–17.

The Development of Hong Kong and Kowloon as Told in Maps by T.R. Tregear & L. Berry, Hong Kong University Press, MacMillan and Co., Ltd., 1959, 31p.

The Hong Kong Region: Its Place in Traditional Chinese Historiography and Principal Events since the Establishment' of Hsin-an County in 1573 by J. Hayes, Journal of the Hong Kong Branch of the Royal Asiatic Society, 1974, 14, 28p.

*The Ruins of War: A Guide to Hong Kong's Battlefields and Wartime Site*s by K.T. Keung & J. Wordie, Joint Publishing (HK) Company Ltd., 1996, 216p.

PART III: Further Reading

1. Geographical

A Geography of China by T. R. Tregear, Aldine Publishing Company, 1965, 342p.

A Geography of Hong Kong edited by T. N. Chiu & C.L. So, Oxford University Press, 1986, 403p.

Geography and the Environment in Southeast Asia by R.D. Hill & J.M. Bray, Hong Kong University Press, 1978, 485p.

Hong Kong and its Geographical Setting by S. G. Davis, Collins, London, 1949, 226p.

Introducing Physical Geography 4th Edition by A. H. Strahler & A. Strahler, John Wiley & Sons Inc., 2005, 752p.

Mapping Hong Kong : A Historical Atlas by H. Empson, Government Information Services, Hong Kong, 1992, 251p.

The Development of Hong Kong and Kowloon as Told in Maps by T.R. Tregear & L. Berry, Hong Kong University Press, 1959, 31p.

2. Geological

A Proposed Revision of the Volcanic Stratigraphy and Related Plutonic Classification of Hong Kong by S.D.G. Campbell & R.J. Sewell, Hong Kong Geologist, 1998, 4, pp. 1–11.

Geological Landscapes of Hong Kong by The Hong Kong Geological Survey, Hong Kong Government Printer, 1998, 61p.

Geology by S. Chernikoff, Worth Publishers, 1995, 593p.

Hong Kong Minerals by C.J. Peng, Hong Kong Urban Council Publication, Hong Kong Government Printer, 1978, 80p.

Second Hutton Symposium on the Origin of Granites and Related Rocks. Excursion C2—Granites of Hong Kong by R. Sewell & R. Langford, Geological Society of Hong Kong Newsletter, 1991, 9(3), pp. 3–28.

Spatial and Temporal Characteristics of Major Faults in Hong Kong by K.W. Lai and R.L. Langford. In: Seismicity in Eastern Asia, edited by R.B. Owen, R.J. Neller & K.W. Lee, Bull. 5 Geological Society of Hong Kong, 1996, pp. 72–84.

3. Climate and Weather

Climate and Weather by P.C. Chin. In: A Geography of Hong Kong, edited by T.N. Chiu & C.L. So, Oxford University Press, 1986, pp. 69–85.

The Urban Climate by W.J. Kyle. In: A Geography of Hong Kong, edited by T.N. Chiu & C.L. So, Oxford University Press, 1986, pp. 86–109.

4. Vegetation and Biology

Checklist of Hong Kong Plants by the Hong Kong Herbarium, Agriculture and Fisheries Department, Bulletin No.1, Hong Kong Government, 1978, 142p.

Hills and Streams, An Ecology of Hong Kong by D. Dudgeon & R. Corlett, Hong Kong University Press. 1994, 234p.

Hong Kong Bamboos by P. But, L, Chia, H, Fung & S. Hu, Hong Kong Urban Council Publication, Hong Kong Government Printer, 1985, 85p.

Hong Kong Freshwater Fishes by M.S. Hay & I.J. Hodgkiss, Hong Kong Urban Council, Hong Kong Government Printer, Hong Kong, 1981, 76p.

Hong Kong Lichens by S. Thrower Hong Kong Urban Council Publication, Hong Kong Government Printer, 1985, 122p.

Hong Kong Seashells by J. Orr. Hong Kong Urban Council Publication, Hong Kong Government Printer, 1985, 122p.

Hong Kong's Wild Places: An Environmental Exploration by E. Stokes, Oxford University Press, 1995, 198p.

Plants in Mangroves by S. Aksornkoae, G.S. Maxwell, S. Havanond & S. Panichsuko, Chalongrat, Bangkok, 1992, 120p.

Soil by C.J. Grant. In: A Geography of Hong Kong, edited by T.N. Chiu & C.L. So, Oxford University Press, 1986, pp. 110–117.

The Green Dragon—Hong Kong's Living Environment by M. Williams & M. Pitts, Green Dragon Publishing, 1994, 142 p.

The Sea Shore Ecology of Hong Kong by B. Morton & J. Morton, Hong Kong University Press, 1983, 350p.

Vegetation by P. C. Catt. In: A Geography of Hong Kong, edited by T.N. Chiu & C.L. So, Oxford University Press, 1986, pp. 118–147.

5. Country Parks and Trails

Across Hong Kong Island: Its Natural Beauty by E. Stokes, Hong Kong Conservation Photography Foundation, 1998, 151p.

Eagle's Nest Nature Trail by Government Information Services, undated, 28p.

Exploring Hong Kong's Countryside by E. Stokes, Hong Kong Tourist Association & Agriculture and Fisheries Department, 1999, 184p.

Magic Walks, The MacLehose Trail and its Surroundings by R. Pearce, the Alternative Press, 1995, 190p.

Selected Walks in Hong Kong by R. Forest & G. Hobbins, A South China Post Publication, Hong Kong, 1979, 96p.

The MacLehose Trail by T. Nutt, C. Bale & T. Ho, Chinese University Press, Hong Kong, 1992, 137p.

6. Historical

Beyond the Metropolis: Villages in Hong Kong edited by P.H. Hase & E. Sinn, The Royal Asiatic Society (Hong Kong Branch), Joint Publishing (HK) Company Limited, 1995, 175p.

From Bondage to Liberation: East Asia 1860–1952 by N. Cameron, Oxford University Press, 1975, 369p.

In The Heart of the Metropolis: Yaumatei and Its People, edited by P.H. Hase, The Royal Asiatic Society (Hong Kong Branch), Joint Publishing (HK) Company Limited, 1999, 181p

Prelude to Hong Kong by A. Coates, Routledge & P. Kegan, 1966, 232p.

The Battle of Hong Kong by G.D. Johnson, After The Battle Magazine, Number 46, 1984, pp. 1–25.

The Guns and Gunners of Hong Kong by D. Rollo, Gunners Roll of Hong Kong, 1991, 10p.

The Hong Kong Story by C. Courtauld & M. Holdsworth, Oxford University Press, 1997, 136p.

The Lasting Honour: The Fall of Hong Kong 1941 by O. Lindsay, Sphere Books Ltd., 1980, 211p.

7. Urban Studies

The New Towns Programme by W.T. Leung. In: A Geography of Hong Kong, edited by T.N. Chiu & C.L. So, Oxford University Press, 1986, pp 251–304.

Twenty Years of New Town Development by the Territory Development Department, Hong Kong Government Printer, 1993, 46p.

Urban Housing and the Residential Environment by L.S.K. Wong, In: A Geography of Hong Kong, edited by T.N. Chiu & C.L. So, Oxford University Press, 1986, pp. 279-304.

Hok Tau, Pat Sin Leng Country Park

Index

A

adits 78, 99, 114, 167
afforestation 23, 85, 97–99, 132–133
agriculture 61
air-fall eruptions 16
alluvium 57
ammonites 8
amphibians and reptiles 206
ancient settlement 61
Anderson Road Quarry 200
animal biodiversity 110
anticline 52, 193
antiform 52
asymmetrical mountains 67

B

back-arc basin 104, 105
badlands 90, 92–93
basalt 22
base level 56, 94–95
batholiths 22
Beacon Hill 173, 186
Beaufort Channel 211
bedding planes 26, 71
beds 26
bioturbation 57, 143
black-eared kite 110
Bluff Head Formation 29, 69, 122
borrow areas (offshore) 214
boulder fields 39
breccia 27, 228
bricks 63
Brides Pool 72
Burmese python 110
butterflies 130

C

calderas 8, 135–136, 154, 222
calderas and columnar joints 136
Cape D'Aguilar Marine Reserve 206
Captain Charles Elliot 189
Castle Peak Range ..38, 84, 90, 92, 94, 96
catchwaters 201
Chek Lap Kok Airport 169–171
Che Kwu Shan Formation 222–223
chi 83
Chinese graves 128
Chinese white dolphin 171–172
Chi Ma Wan Granite 218
Chi Ma Wan Peninsula 151, 156–158
                                                          160
clans 61
Clear Water Bay Formation 135, 224
cliffs (sea) 45
Closed Frontier Area 77
coal 27
coastal environments 137–145
coastal processes
    beach boulders 160
    beach deposits 139
    beach processes 141
    beach terminology 139
    breakers 140–141
    coral coasts 144–145
    deltas 142–144
    delta sediments 143
    fluid density 143
    headland erosion 138
    joint slots 138
    longshore drift 141
    artificial shores 160
    mangroves and muddy flats 142
    pebble beach (Tong Fuk) 163
    planar cross beds 163
    ripples 163
    rip currents 141
    sea arches 139
    sea caves 139
    sea stacks 139
    swash and backwash 141, 163
    tidal flats 161–162
    tombolo 76, 218–219
    wavelength 141
    wave refraction 138
coastlines 42
College of Medicine 197
colluvium 91, 158
compression 35
conglomerate 27, 68, 71–72, 122, 227
contaminated mud 215

coquinas 27, 143
corals 81, 144–45, 206, 209, 225
corestones 39, 178–179
cormorants 216
country parks 112
crested goshawk 110
Crooked Harbour 228
Crooked Island (*See* Kat O)
cuesta 18–19, 25

D

debris flows 158–159
Deep Bay 41, 49, 61
deforestation 60–61
detrital 27
diatoms 215
differential erosion 30–31, 50, 71, 89
dip angle 18
dip slopes 70–71, 227–228
Discovery Bay 151
dolphins 171–172, 209, 215, 218
dolphin habitat problems 172
Double Haven 81, 229
dykes .. 22, 137, 152, 154–155, 217–218
                                                          221
dykes, feldsparphyric 155

E

earthquake risks 108
Earth movements 25, 35, 52
East Lamma Channel 33, 211, 214
erosion 89, 148, 200
escarpments 18
eutaxite 193
eutrophication 215
extrusive igneous rocks 15

F

Fan Lau 164
Fan Lau Granite 156
faults
     Lantau 36
     orientation 33
     rock weakening 19, 194, 225
     thrust fault (Double Island) 229
     thrust fault (Tiu Tang Lung) 70

Tolo Channel Fault Zone ... 107–108
     fault types 35
Fei Ngo Shan 119, 173
Feng Shui 82–83
Feng Shui woodland 83
fissure eruptions 122, 135
floral diversity 5
folds 52
footpaths 148
foraminifera 215
formations 28–29
forts 164–165, 224
Fung Wong Shan 149

G

gei wei 41, 55, 64
geological dating 6
geological history 7–10, 29
Gin Drinker's Bay 101
glacials and interglacials 125
golf courses 148, 226
Government Hill 196
Governor's Beach 229
granite 22, 86–87, 107, 154, 156–157
                                                 175–176, 192, 218
granite weathering 92
granitic rock suites 87, 107
granodiorite 22, 86–87
graphite schist 27, 216
Grassy Hill 101
gullies 90, 92–93

H

Hakka 61–62, 111, 127, 145, 151
half-life 6
haze 199
High Island Formation 135, 222, 224
hill fires 80
historical change
     City of Victoria 195–198
     Hong Kong Island 204
     Kowloon 174
     northwest New Territories 61
Hoi Ha Wan Marine Park 144
Hoklo 62, 151
Hong Kong Island 189–206

Hong Kong minerals 100
Hong Kong and Shanghai Bank 196
human impacts (offshore) 96–97, 214
human occupation 61, 82, 109, 113
                        127, 145–146, 164, 204

I
Ice Age 125
ignimbrite 15
insects 134
intrusive igneous rocks 22
islands
    Ap Chau 228–229
    Basalt Island 224–226
    Beaufort Island 220
    Bluff Island 224–226
    Brothers Islands 170, 216
    Chek Chau 227
    Chek Lap Kok 169–171, 216
    Cheung Chau 218, 220
    Dangan Islands 213
    Double Island 229
    East Brother Island 216
    Hei Ling Chau 218, 220
    Jin Island 226
    Kat O 227–229
    Kau Sai Chau 226–227
    Kau Yi Chau 218
    Lung Kwu Chau 215–216
    Ninepin Islands 222–223
    Peng Chau 218
    Ping Chau 74, 81, 227–229
    Po Toi 220, 222
    See Chau 226
    Sharp Island 224, 226–227
    Sha Chau 215, 216
    Shek Kwu Chau 218
    Shelter Island 224, 226
    Siu A Chau 217
    Soko Islands 217–218
    Sunshine Island 218, 220
    Tai A Chau 217, 218
    Tai Tau Chau 226–227
    Tap Mun 227–229
    Town Island 224–225
    Tree Island 215–216
    Tung Lung Island 223–224
    Waglan Island 221–222
    Wang Chau 224
    West Brother Island 216
    Yim Tin Tsai 226–227
isostacy 125

J
James Hutton 5–6
joints
    columnar 37, 45, 136
                        222–223, 225
    erosion 71
    rock weakening 154, 225
    sheeting 37, 221
    structure and types 36–38
    tectonic 37

K
Kadoorie Farm 106
Kai Sai Chau Volcanic Group 122
Kam Tin 34, 63
Kam Tin Valley 55
kaolin 34, 167
Kap Man Hang 17, 147
Kat O Formation 29, 228
Kowloon and Canton Railway 188
Kiu Tsui Country Park 227
koel 110
Kop Tong 82
Kowloon Granite 122, 176
Kowloon Walled City 174–175
Kwun Yam 106

L
lahars 9, 106, 153
Lai Chi Chong Formation 133
Lai Chi Wo 82
Lamma 191
Lam Tei Quarry 200
Lam Tsuen Valley 33, 54, 56
landfills 129
landscapes (coastal) 41–46
landscapes (granitic) 21–23
landscapes (joints & faults) 33–39
landscapes (sedimentary) 25–31

Index

landscapes (volcanic) 13–19
landscape colours 28
    carbon 28
    copper 28
    iron 28
    manganese 28
landscape photographs
    Aberdeen Reservoirs 193
    Anderson Road Quarry 200
    Ap Chau 24, 70, 228
    Basalt Island 45, 226
    Beaufort Island 221
    Bluff Head 28, 69
    Bluff Head Ridge 25, 67
    Bluff Island 19, 45, 208, 225
    boundary stones 195
    Brides Pool 72
    Buffalo Hill .... iii, 245
    Byewash Reservoir 185
    Castle Peak Range 23, 84, 90, 92, 96
    Castle Peak Range waterfalls 94
    catchwater (Hong Kong Island) .. 201
    Chek Chau 70, 227
    Chek Keng 144
    Chek Lap Kok reclamation ..170–171
    Cheung Chau 219, 220
    Cheung Sha Beach 160
    Cheung Uk 253
    Chi Ma Wan Peninsula 157, 160
    Closed Frontier Area 77
    columnar jointing 135, 137, 147, 223

Crescent Island 229
Crooked Harbour 74
dolphins 171, 172, 216
Double Island 229
Dragon Pool 73
Dragon's Back 190
Fanling 58
Fan Lau 43, 157
Fan Lau Fort 164
Fei Ngo Shan 121, 183
Governor's Beach 229
Grassy Hill 103
Hei Ling Chau 219
High Island Reservoir 146
Hok Tau 241
Hong Kong Island 190–191, 193–194
Kai Kung Leng 34, 48, 53
Kam Shan Country Park 111
Kam Shan (reservoirs) 184
Kam Tin (floodplain) 54
Kap Man Hang 16, 229
Kat O 46, 74, 229
Kop Tong 82
Kowloon 173, 180, 234
Ko Lau Wan 146
Kwun Yam 34, 106
Lai Chi Wo 82
Lamma 191, 199, 222, 238
Lamma (Sham Wan) 205
Lamma (slump) 202
Lam Tsuen Valley 59
Lam Uk 140

Buffalo Hill, Ma On Shan Country Park

245

Lantau Peak 149, 156, 167
Legislative Council (building) 197
Lin Ma Hang (bats) 78
Lin Ma Hang (mine) 78, 79
Lion Rock .... v, 47, 174, 176, 186
Long Harbour 142
Long Ke 140
Long Ke Tsai 19
Lower Shing Mun Reservoir 111
Lui Ta Shek (panorama) 132
Luk Keng 249
Lung Kwu Chau 216
Lung Kwu Tan 20, 23
Lung Tsai Ng Yuen Gardens 168
macaques 110, 188
Mai Po 32, 41, 49, 55, 62
64, 162
mangroves 32, 160
Ma On Shan 11–12, 14, 117
Ma On Shan Country Park ..119–120
Ma On Shan (landslips) 129
Ma Shi Chau 76
Ma Wan Chung 150
Mirror Pool 73
Murray House 196
Nei Lak Shan 172
Ngau Ngak Shan 123
Nga Ying Shan 149
Ngong Ping 167
North Lantau Expressway 169
North Ninepin Island 223
Pat Sin Leng 17, 80
Pearl River Estuary (islands) 208
Peng Chau 219–220
pill boxes 187, 205
Ping Chau 30, 74, 228
Ping Fung Shan 18, 70
Plover Cove Reservoir 80
Plover Cove Country Park 230
Pok Fu Lam Reservoir 201
Pok Fu Lam Valley 194
Po Pin Chau 137
Po Toi 221–222
Queen's Rd. 198
Robin's Nest (panorama) 77
Sai Kung coastal erosion 139

Sai Kung East Country Park 135
Sai Kung (islands) 126, 148
Sai Wan 140, 148
Sharp Island 227
Sharp Peak 14, 131, 148
Sha Chau 171
Sha Lo Tung 68
Sha Tau Kok 75
Sha Tin 103, 109, 115
Shek Nga Shan 122
Shek O 44
Shing Mun Gorge 113
Shing Mun Redoubt (tunnels) ... 188
shotcrete 202
Shui Hau tidal flat 161–162
Shui Lo Cho 153
silver mine (Mui Wo) 167
Siu A Chau 217
slopes (mist nets) 204
slopes (stone pitching) 204
Soko Islands 207, 217
South Ninepin Island 223
squatter communities 113
St. John's (cathedral) 196
Stone Trail 16, 236
stream erosion 200
Sunset Peak 156
Sunshine Island 219
Tai Hang Valley 54
Tai Lam Country Park 85, 96
Tai Lam Reservoir 230
Tai Long Wan 1, 40, 140
Tai Mo Shan 101, 103, 106
Tai O 151, 166
Tai Po 109
Tai Tam Tuk Reservoir 201
Tai To Yan 34, 53
Tap Mun 227
tea terraces 109
Tian Tan Buddha 168
Tin Hau temple 222
Tiu Tang Lung 70, 74
Tolo Channel 75
Town Island 225
Tung Chung 232
trawling 214

Tsim Chau 46
Tsing Ma Bridge 169
Tuen Mun 96
Tung Lung Island 224
Turret Hill 107
Two International Finance Centre 198
University of Hong Kong 197
Victoria Harbour 44, 173, 189
Waglan Island 221
Wang Leng 70
Waterfall Bay 203
weathered profiles 178–179
white-bellied sea eagles 220
Wong Chuk Shan 127
Wu Kau Tang 82
Yam O 166, 169
Yi O 150
landslides 158–159, 202
land animals 5
land area 4
land sales 195–196
Lantau 36, 42, 87, 149–172
Lantau (country parks) 152
Lantau (dyke swarm) 155
Lantau Granite 153, 192
Lantau Peak 149
Lantau (vegetation) 152
Lantau Volcanic Group 153
Lan Tau Pai 46
Lau Fau Shan 49
lead-210 6
Lead Mine Pass 101
Legislative Council (building) 197
Lei Yue Mun Gap 33
leopard cat 110
levee 57
Lianhuashan Fault Zone 34
limestone 7, 27
Lin Fa Shan 99
Lin Ma Hang 77–79
Lion Rock 173
Lok Ma Chau Formation 29
Long Harbour 131
Long Harbour Formation 135
Luk Chau Shan 38

M

macaques 110, 188, 206
magma 14–15, 22, 86, 105
                              175, 192, 220
magma chambers 8, 86, 107, 121, 218
Mai Po 49, 55, 62–64
mangroves 41, 49, 64, 142
Mang Kung Uk Formation 133
maps (geology)
    central New Territories 104
    eastern New Territories 133
    Hong Kong 13, 21, 25
    Hong Kong Island and Lamma .. 192
    islands (off eastern Lantau) 218
    islands (Mirs Bay) 227
    islands (Po Toi group) 220
    islands (Rocky Harbour
       and Port Shelter) 224
    islands (Soko) 217
    islands (southeastern) 222
    islands (western) 215
    Kowloon and Lion Rock 175
    Lantau 153
    Ma On Shan (iron mine) 123
    Needle Hill 114
    northeastern New Territories 69
    northwestern New Territories 50
    southeastern New Territories 119
    Shui Hau (tidal flat) 161
    Victoria Peak 193
    western New Territories 86
maps (miscellaneous)
    calderas 136
    Cape D'Aguilar Marine Reserve .. 206
    Castle Peak Range 94
    central New Territories 101
    city of Victoria 195
    coastal change (Mai Po) 55
    coastal change (northwestern
       New Territories) 63
    coastal erosion 42
    eastern New Territories 131
    faults (Kowloon reservoirs) 184
    fault valleys (Hong Kong Island) .. 194
    Hong Kong Island 189
    human impacts (sea floor) 214
    islands (Hong Kong) 207

Japanese invasion 187
Kam Tin River 66
Kowloon and Lion Rock (hills) .. 173
Kowloon (pluton) 176
Kowloon (weathering front) 179
Lantau 149
Lantau (country parks) 152
late Quaternary 125
late Quaternary (shorelines) 212
late Quaternary (vegetation) 60
Ma On Shan (iron mine) 124
northeastern New Territories 67
Ng Tung River 66
north Lantau (marine problems) .. 172
northwestern New Territories 48
ocean currents 209
Ping Chau Marine Park 81
Pleistocene (rivers) 170, 212
reclamations (Victoria Harbour) .. 182
reclamations (western New
  Territories) 96
Scheduled Areas 118
sea floor bathymmetry 211
sea Levels (Sai Kung) 126
Sha Tin Valley (development) 116
Shing Mun Redoubt 186
southeastern New Territories 119
tidal pathways 211
Tolo Channel (fault) 108
Tuen Mun (weathering front) 90
weather 132
western New Territories 84
Yan Chau Tong Marine Park 81
marble 7, 27, 50, 116, 123
mariculture 227, 229
marine parks 81
marine salinity gradients 208–209
mass movements 158–159
Ma On Shan 119, 123
Ma On Shan Formation 29
Ma On Shan (iron mine) 123, 124
metamorphic rocks 50–52
migrating volcanoes 105
minerals
  biotite 22, 100
  calcite 100
  feldspar 22, 86, 100, 154
  fluorite 114
  galena 78, 100, 114, 166
  graphite 166–167, 216
  haematite 22
  hornblende 100
  limonite 28
  magnetite 123
  molybdenite 100, 114
  pyrite 114
  quartz 22, 78, 86, 100, 154
  silver 166
  sphalerite 100
  wolframite 99–100, 114, 167
mining
  kaolin 167
  Lin Fa Shan (tungsten, adits) 99
  Lin Ma Hang (lead) 78–79
  Ma On Shan (iron) 124
  Needle Hill (tungsten) 114
  Silver Mine Bay (silver) 166
  West Brother Island (graphite) .. 216
mining (landscape scars) 114
Mirs Bay 227
mist nets 204
monzonite 22
mountain streams 94
Mt. Butler Granite 122, 176, 192
Mt. Butler Quarry 200
Mt. Stenhouse 191
mudstone 27, 71
mud flats 41
Mui Wo 151
Murray House 196

N
Nautilus 8
Needle Hill 101, 114
Needle Hill Granite 107, 114, 121
                                       154, 175
Nei Lak Shan 156, 172
New Towns 115
Ngau Ngak Shan 119, 123
Ngo Mei Chau Formation 228
Ng Tung Chai Waterfalls 102
nuée ardente 15

nullahs 66

O

ocean currents 209
ocean waves 19
offshore buried landscapes 212
oldest lighthouse 222
oysters 49

P

Pat Sin Leng 17, 67
Pat Sin Leng Formation 28–29, 69–70
paved trackways 127
Peak Tram 197
pearls 229
Pearl River 208
photographs (aerial)
    Kowloon and Sha Tin 3
    Lin Fa Shan 99
photographs (historical)
    Kowloon Walled City .. 174–175, 180
    Sheung Shui (floodplains) 65
    Tide Cove (Sha Tin) 117

photographs (marine)
    anemones 226
    corals (Hoi Ha Wan) 144
    corals (Ping Chau) 81, 209
    corals (Soko Islands) 218
    diatoms 215
    foraminifera 215
    mudskippers 160
    ray (Soko Islands) 218
    starfish (Soko Islands) 218
photographs (miscellaneous)
    abandoned villages 68
    bedding planes 26, 71
    Kam Tin (nullah) 66
    Kam Tin River (point bar) 66
    mangrove leaves 64
    potholes 71
photographs (rocks and sediments)
    basalt 22
    basaltic andesite 154
    beach sands 163
    biotite granite 176
    coal 27

Luk Keng, Plover Cove Country Park

conglomerate 18, 72
gneiss 51
granite 22, 87
granite (Po Toi) 221
granodiorite 22
graphite schist 27, 216
ignimbrite 15
limestone 27
marble 51
monzonite 22
pebbles (Tong Fuk coast) 163
pebbles (and Tai O Formation) .. 152
phyllite 51
porphyry 154
pumice tuffs 106
rhyodacite 155
rhyolite 22, 121, 135, 155
rhyolite (feldsparphyric) 22
rhyolitic hyaloclastite 106
sand 27
sandstone (Bluff Head) 28
sandstone (Chek Chau) 30
sandstone (Ping Fung Shan) 26
sandy ripples 163
schist 51
shale 51, 72
shale (Sai Kung Country Park) 28
siltstone (Ma Shi Chau) 30
siltstone (Ping Chau) 30
siltstone (Shek Uk Shan) 26
siltstone and quartz 31
tuff 15, 18, 135
pill boxes 187
Ping Chau Formation 29, 228
Ping Fung Shan 18, 26, 30
Ping Shan 63
pirates and forts 165
plantation trees 98
plate boundary 14, 105
plate tectonics 88
Pleistocene rivers 212
Plover Cove 67
plunge pool 72, 94
plutons 22, 176
pollution 199, 234
population movements 61–62

population pressures 113
Port Island Formation 28–29, 227
Port Shelter 224
potholes 71
Po Lin (monastery) 168
Po Shan Road (landslide) 202
Po Toi Granite 220
public housing 113
Punti 62, 127

Q

quarries 147, 199–200
quartz veins 88–89

R

radioactive elements 6
reclamations
    Central 197
    Chek Lap Kok 169–171
    coastal 214
    Kowloon 180, 182
    northwest New Territories 63–64
    Sha Tin 115–116
red tides 215
reefs 81
religion and Lantau 167–168
Repulse Bay Volcanic Group 122
reservoirs
    Aberdeen Reservoir 201
    Byewash Reservoir 185
    High Island Reservoir 146
    Hong Kong 111
    Kowloon 184
    Kowloon Reservoir 184
    Plover Cove Reservoir 80
    Pok Fu Lam Reservoir 201
    Reception Reservoir 185
    Shek Lei Pui Reservoir 185
    Shek Pik Reservoir 152, 156
    Shing Mun Reservoir 111
    Tai Lam Chung Reservoir 97
    Tai Tam Reservoir 201
    Wong Nei Chung Reservoir 201
rhyodacite 155, 218
rhyolite ... 14–15, 22, 122, 154–155, 221
ria 131

rills 93
rivers
    alluvium 57
    base level 56, 59
    floodplains 54, 56, 64–66
    levee 57–58
    maturity 54
    nullahs 66
    oxbow lake 56
    point bar 56–57, 66
    rejuvenation 56
    river processes 54–59
    terraces 58–59
Robin's Nest 77
rock carvings (Nam Tam Wan) 222
rock platforms 46, 76

S

Sai Lau Kong Formation 106
salt pans (Lantau) 166
sandstone 27, 68, 71, 122, 227, 229
sand dunes 9
Scheduled Areas 118
sea caves 45, 139
sea level change ... 10, 125–126, 212–213
sedimentary formations 28–29
sedimentary rock colours 28
seismic profiles 213
Sha Chau and
Lung Kwu Chau Marine Park 171, 215
Sham Chun River 41
Sham Wan 206
Sha Lo Wan 167
Sha Tau Kok 18, 33, 63, 75
Sha Tin 101, 115–117
Sha Tin Granite ..107–108, 121, 154, 175
Sha Tin Valley 33, 108
sheeting joints (*See* joints, sheeting)
Shek Kip Mei (squatter village) 113
Shek Kong 34
Shek O Quarry 200
Shek Uk Shan 26
Shing Mun Formation 106, 133, 153
Shing Mun Redoubt 186, 188
Shing Mun Valley 101
shotcrete 202

Shui Chuen O Granite 107, 121, 175
Shui Hau 161
Shui Ngau Shan 119
silica 22
sills 22
siltstone 27, 68, 227
Sir Murray MacLehose 112–113
skarn 123
slopes (stability and maintenance) 202
slumps 129, 158–159
soil pipes 93
Sok Kwu Wan Quarry 199–200
sombrero islands 76
South Lamma Granite 192
spheroidal weathering 39
spring sapping 93
stacks 45, 138
Starling Inlet 74–75
stone pitching 204
Strike-slip fault (*See* faults, types)
subduction 14, 88, 105
subtidal platforms 74
Sunset Peak 149
superposition 6
supratidal platforms 76
syncline 52, 193
synform 52

T

Tai Chau 46
Tai Lam Country Park 96
Tai Lam Granite 87
Tai Long Tsui 138
Tai Long Wan 47, 140
Tai Mo Shan 101
Tai Mo Shan Formation 106, 228
Tai O 151
Tai O Formation 152
Tai Po Granodiorite 107, 192
Tai Po Kau 116
Tai Po Pass 108
Tanka 62
Tate's Cairn 173
tea terraces 109
tension (crustal) 35
terrace walls 127

thrust fault (*See* faults, types)
Tian Tan Buddha 168
tidal channels 209–212
tides 209–210
Tin Hau 216, 218
Tolo Channel 33, 74, 108
Tolo Channel Formation 29
Tolo Harbour Formation 29
tors 23
toxic sea floor pits 215
Trappist Haven Monastery (Lantau) .. 168
tropical hardwoods (Yam O) 166
tropical rainforests 60, 129
Tsing Ma Bridge 169
Tsing Shan Granite 87
Tsuen Wan 101
Tsuen Wan Volcanic Group 104, 106
Tuen Mun 96
tuffs 15–16, 19, 68, 87, 121, 193
                                    222, 227–229
Tung Chung 150, 232
Turret Hill 147
turtle nesting 206
Two International Finance Centre 198

U

unconformity 152
uncontaminated mud dumping 215
uniformitarianism 5–6
University of Hong Kong 197
urbanisation and streams 200
urban canyons 198
urban climates 198
urban development 182
urban geological mapping 181
urban geology 183
urban pollution 198–199
Urmston Rd. 214

V

valley deltas and progradation 126
Victoria Peak 189
Vietnamese refugees 217
volcanic arc 105
volcanic ash 14
volcanic bomb 15

volcanic groups 87
volcanic rocks 13
volcanoes 86

W

Wang Leng 67
waterfalls 16, 37, 71–73, 94–95, 102
                                    147, 153, 202–203
weather 132
weathered granite 23, 92
weathering 154, 176, 179
weathering front 177–179
weepholes 202
wetlands 41, 49, 55
white-bellied sea eagle 110
wildlife 77, 109–110, 116
                                    133, 160, 188, 205–206
wildlife (marine) 218, 226
wild boar 110
Wong Leng 67
World War II 186–187
Wu Kau Tang 82

Y

Yan Chau Tong 81
Yim Tin Tsai Formation 104, 193
Yuen Long 63
Yuen Long Formation 29
Yuen Long Plain 54, 55
Yuen Tsuen (ancient trail) 96

Cheung Uk, Sha Lo Tung
Pat Sin Leng Country Park